Fibre Optics

Fibre Optics
Theory and Applications

Serge Ungar
Aerospatiale, Cannes, France

Translated by
John C. C. Nelson
University of Leeds, UK

JOHN WILEY & SONS
Chichester · New York · Brisbane · Toronto · Singapore

First published under the title 'Fibres Optiques: Théorie et applications'
by Éditions Bordas, © BORDAS, PARIS 1989

Copyright © 1990 by John Wiley & Sons Ltd.
Baffins Lane, Chichester
West Sussex PO19 1UD, England

Other Wiley Editorial Offices

John Wiley & Sons, Inc., 605 Third Avenue,
New York, NY 10158-0012, USA

Jacaranda Wiley Ltd, G.P.O. Box 859, Brisbane,
Queensland 4001, Australia

John Wiley & Sons (Canada) Ltd, 22 Worcester Road,
Rexdale, Ontario M9W 1L1, Canada

John Wiley & Sons (SEA) Pte Ltd, 37 Jalan Pemimpin 05-04,
Block B, Union Industrial Building, Singapore 2057

Library of Congress Cataloging-in-Publication Data:

Ungar, Serge.
    [Fibres optiques. English]
    Fibre optics : theory & applications / Serge Ungar ; translated by
John C. C. Nelson.
        p.   cm.
    Translation of: Fibres optiques.
    Includes bibliographical references.
    ISBN 0 471 92758 9
    1. Fiber optics.   2. Fiber optics—Industrial applications.
I. Title.
    TA1800.U5213   1990
    621.36' 92—dc20
                                                    90-35079
                                                    CIP

British Library Cataloguing in Publication Data:

Ungar, Serge
    Fibre optics.
    1. Fibre optics
    I. Title
    621.3692

ISBN 0 471 92758 9

Typeset by Mathematical Composition Setters Ltd, Salisbury
Printed in Great Britain by Biddles Ltd, Guildford.

CONTENTS

Preface   ix

1. The optical fibre   1

1. Definition   1
2. The numerical aperture   2
3. The refractive index profile   3
4. Monomode and multimode optical fibres   4
5. The absorption spectrum of a fibre   5
6. Comparison of multimode and monomode optical fibres   5

2. Graded index multimode optical fibres   7

1. Introduction   7
2. Definition of the refractive index profile   7
3. The wave equation   8
4. Analytic solution   9
5. Solution by the WBKJ method   10
6. Interpretation of the wave function   11
7. The influence of the propagation constant   13
8. The number of modes in a fibre   16
9. Dispersion in optical fibres   20
10. Determination of the optimum refractive index profile   23
11. Variation of refractive index profile   25
12. The influence of coupling between modes   30
13. The frequency response of the fibre   31
14. The effective numerical aperture   33
15. Examples of multimode fibres   34

3. Monomode optical fibres   37

1. Introduction   37
2. Definition and solution of the scalar wave equation   37
3. Application to a step index monomode optical fibre   39

    4. Cutoff frequency   45
    5. Degeneracy of modes   47
    6. Monomode propagation conditions   48
    7. The confinement factor   48
    8. Spectral attenuation   52
    9. Dispersion in monomode fibres   52
   10. Multidielectric structures   53
   11. The performance of monomode optical fibres   54

4. Fabrication of optical fibres   55

    1. The principles of fabrication   55
    2. Fabrication of preforms   56
    3. The influence of fabrication materials   62
    4. Control of the fabrication process   64

5. Characterization of optical fibres   67

    1. Introduction   67
    2. Numerical aperture   67
    3. Measurement of refractive index profile   68
    4. Measurement of losses in an optical fibre   75
    5. Measurement of the frequency characteristics of optical fibres   82
    6. Observation of mode groups   85
    7. Backscattering methods   89
    8. Measurements specific to monomode optical fibres   93

6. Fibre optic cables   97

    1. Introduction   97
    2. Mechanical properties and endurance of optical fibres   98
    3. Protection of optical fibres before cabling   99
    4. Materials used in cables   99
    5. The structure of optical cables   101
    6. The performance of cables   104
    7. The parameters of optical cables   105
    8. Installation of optical fibre cables   106

7. Coupling of optical fibres   109

    1. Introduction   109
    2. Intrinsic parameters which cause losses   109
    3. Extrinsic parameters   112
    4. Preparation of the fibre ends   117

5. Principles of coupling   118
6. Connection by welding (fusion)   119
7. Splices   119
8. Connectors   120
9. Connection specification   120
10. Passive couplers   121
11. Active couplers   124
12. Miscellaneous coupling systems   124
13. Choice of coupling   124

8. The light-emitting diode (LED)   125

1. The principle of light emission in a semiconductor   125
2. The p–n junction   126
3. Structures   129
4. The efficiency of a diode   131
5. Flux and luminous intensity   137
6. Optical parameters of the diode   141
7. Electrical and thermal properties of the diode   142
8. Electrical parameters of the diode   145
9. Typical examples   150
10. The bandwidth of a light-emitting diode   152
11. Examples of light-emitting diodes   154

9. Laser diodes   155

1. Introduction   155
2. The amplifying medium   155
3. The resonant cavity   159
4. Laser diodes   163
5. Laser diode optical heads   170
6. Laser diode noise   173
7. Characteristics of laser diodes   173

10. The receiver   175

1. Photon-electron conversion   175
2. The principle of photodetectors   179
3. Speed   180
4. The pin photodiode   181
5. Structure   184
6. The avalanche photodiode (APD)   185
7. Equivalent electronic circuit of a pin diode   187
8. Noise   188

  9. Signal-to-noise ratio   189
10. Additional characteristics   190
11. Specific circuits   191
12. Example circuits   192
13. Bandwidth of a photodiode   194
14. PIN diode characteristics   194

11. Applications of optical fibres   **197**

  1. Optical fibre telecommunications   197
  2. Fibre optic transducers   212
  3. Imaging   219

Bibliography   **221**
Index   **223**

PREFACE

This work, devoted to optical fibres, is primarily concerned with industrial applications in order to acquaint engineers with the technology and the associated optics and electronics. No particular knowledge is required with the exception of a sound mathematical background.

Three industrial aspects of fibres are considered; these are fabrication, interfacing and applications. Each of these requires knowledge of the other two so it seemed appropriate to present them together in the same work.

The first three chapters are fundamental since they present the fibre from a theoretical point of view; this enables all its characteristic parameters to be defined together with the effect of external conditions on these parameters.

Chapters 4, 5 and 6 describe the fabrication of fibres and the monitoring techniques used to check the parameter values.

Chapters 7, 8, 9 and 10 describe interfaces and telecommunications applications, particularly by developing the transmitting, receiving and coupling aspects.

The last chapter presents optical fibre applications in an industrial environment with emphasis on transmission and an explanation of the basic methodology required for evaluation of fibre optic links. A large part of this chapter is also devoted to fibre optic transducers.

A concise bibliography is given at the end of the work to guide the interested reader in developing certain points to greater depth. It should be mentioned that the theory presented has been developed in detail by the engineers of the Thomson Company and particularly by the team of M. Spitz of the Thomson Central Research Laboratory.

1

THE OPTICAL FIBRE

1. DEFINITION

An optical fibre is a light guide governed by Snell's law; it consists of a core in which light waves propagate by reflection from an optical cladding of lower refractive index (Fig. 1.1).

Snell's law defines the passage from a medium of refractive index n_1 to a medium of refractive index n_2 by a light ray having an angle of incidence i (Fig. 1.2).

$$n_1 \cdot \sin(i) = n_2 \cdot \sin(r) \tag{1}$$

where r is the angle of the refracted ray in medium 2 (Fig. 1.2).

There is a value of the angle of incidence for which the wave is reflected at the medium interface (the Brewster angle),

$$\sin(i_M) = n_2/n_1 \quad \text{if} \quad n_2 < n_1 \tag{2}$$

which requires that the second medium has a refractive index less than that of the first. If the angle of incidence i is greater than the limiting angle, the light wave is reflected; this is the case for the optical fibre.

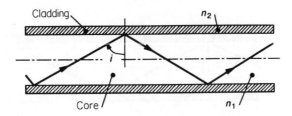

Figure 1.1 Diagram of an optical fibre

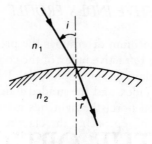

Figure 1.2   Snell's law

2. THE NUMERICAL APERTURE

The numerical aperture defines the maximum angle which the incident beam must make to ensure that it propagates in the fibre (Fig. 1.3).

$$n_0 \cdot \sin(i_0) = n_1 \cdot \sin(i) \tag{3}$$

$$n_1 \cdot \sin(i_1) = n_2 \cdot \sin(i_2) \tag{4}$$

For there to be total internal reflection, it is necessary that $i_2 = 90°$. Hence

$$\sin(i_1) = n_2/n_1 \tag{5}$$

Since $i_1 = 90° - i$

$$\sin(i_0) = \sin(i) \cdot n_1/n_0 = (n_1^2 - n_2^2)^{1/2}/n_0 \tag{6}$$

the numerical aperture NA is defined by

$$NA = (n_1^2 - n_2^2)^{1/2}/n_0$$

The numerical aperture of an optical fibre is usually of the order of 0.2 to 0.3; the greater the numerical aperture, the greater the luminous power injected into the fibre.

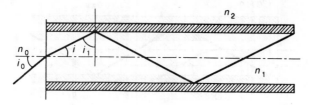

Figure 1.3   Definition of numerical aperture

3. THE REFRACTIVE INDEX PROFILE

The refractive index of a medium determines the propagation of light in the waveguide. In fact, it is the refractive index of the core n_1 with respect to that of the cladding n_2 which plays an important role. Two classes of profile can be distinguished, the step index and the graded index (Figs 1.4 and 1.5).

In a graded index optical fibre, the light has a trajectory which becomes more and more curved as it approaches the cladding (Figs 1.6 and 1.7).

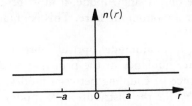

Figure 1.4   Step index profile

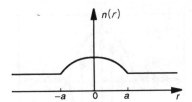

Figure 1.5   Graded index profile

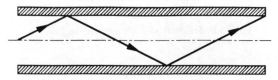

Figure 1.6   Propagation in a step index fibre

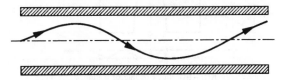

Figure 1.7   Propagation in a graded index fibre

4. MONOMODE AND MULTIMODE OPTICAL FIBRES

An optical fibre is a rotationally symmetrical waveguide which, as with all waveguides, possesses propagation modes. By definition, stationary modes appear in a resonant cavity with intensity nodes at the walls (Fig. 1.8).

The anti-node of maximum amplitude is at the centre of the cavity and the amplitudes diminish towards the exterior. The number of nodes and anti-nodes is the same and, in a multimode optical fibre, there are more than 300 modes.

For a cavity to have only a single mode, it must be at least the size of this mode, that is 6 μm for a monomode fibre. This is obtained with a particular index profile (Fig. 1.9).

The advantage of a monomode optical fibre in comparison with a multimode one is better guidance of the principal mode and a reduction of losses due to the absence of coupling between modes. A distance of around 400 metres is necessary for the distribution of modes in an optical fibre to stabilize.

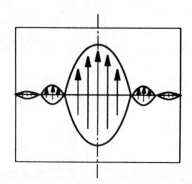

Figure 1.8 Stationary waves in a rectangular guide

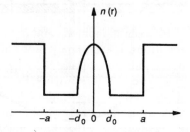

Figure 1.9 Refractive index profile of a monomode fibre

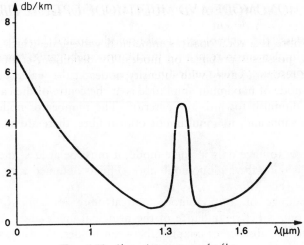

Figure 1.10   Absorption spectrum of a fibre

5. THE ABSORPTION SPECTRUM OF A FIBRE

A fibre has two absorption minima with respect to wavelength and these are at 1300 nm and 1500 nm, in the near infra-red.

At present, the window most used for fibre optic telecommunication is that at 1300 nm. A wavelength of 860 nm is frequently used; although of inferior loss performance it has the advantage of a vestigial spectrum which is visible to the eye.

6. COMPARISON OF MULTIMODE AND MONOMODE OPTICAL FIBRES

The two main differences between monomode and multimode fibres relate to their bandwidth and attenuations. Usually, multimode fibres have bandwidths of 400 MHz/km while monomode fibres easily exceed GHz/km. Similarly, the attenuation of a multimode fibre is 3 dB/km while that of a monomode fibre is 0.4 dB/km. This means that between monomode and multimode fibres there is a factor of 5–10 in performance. By way of example, multimode links have been established with capacities of 144 Mbit/s and repeaters every 20 km while monomode links have been established with capacities of 560 Mbit/s and repeaters every 40 km.

Nevertheless, the technology of monomode optical fibres is much more difficult to implement in respect of installation, debugging, transmitting and receiving components and connectors.

2

GRADED INDEX MULTIMODE OPTICAL FIBRES

1. INTRODUCTION

Two different theoretical approaches are available for the analysis of graded index multimode optical fibres (GRIN). One is rigorous and uses Maxwell's theory in a modal approach to GRIN fibres; the other is approximate and determines mode coupling, thereby permitting intermodal dispersion to be calculated.

2. DEFINITION OF THE REFRACTIVE INDEX PROFILE

The refractive index profile is assumed to be cylindrically symmetrical and to depend only on the radial distance r from the centre of the core to the cladding. Let

$$n_1 = \text{Max}(n(r)) = \text{Max}(n(R)) \tag{1}$$

where $R = r/a$ is the normalized radius.

$$n^2(R) = \begin{cases} n_1^2(1 - 2\Delta R^\alpha) & R \leqslant 1 \\ n_1^2(1 - 2\Delta) = n_2^2 & R \geqslant 1 \end{cases} \tag{2}$$

where

$$\Delta = (n_1^2 - n_2^2)/2n_1^2 \qquad n_1 - n_2 \neq 10^{-2} \tag{3}$$

and α is the refractive index profile parameter.

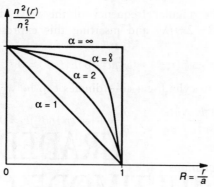

Figure 2.1   Refractive index profile

This configuration allows a transition from a triangular profile ($\alpha = 1$) to a step index ($\alpha = \infty$) while passing through a parabolic profile ($\alpha = 2$), as shown in Fig. 2.1.

3. THE WAVE EQUATION

3.1. The wave equation

The propagation of an electromagnetic wave is governed by Maxwell's equations which for isotropic media can be written as follows:

$$\text{curl}(E) = -\mu \frac{\partial H}{\partial t} \tag{4}$$

$$\text{curl}(H) = \varepsilon \frac{\partial E}{\partial t} \tag{5}$$

$$\text{div}(D) = 0 \tag{6}$$

$$\text{div}(B) = 0 \tag{7}$$

where E is the electric field of the wave, H is the magnetic field, D and B are the electric and magnetic inductions, ε is the permittivity of the medium and μ is its permeability. Combining equations (4) and (5),

$$\Delta E = -\mu\varepsilon \frac{\partial^2 E}{\partial t^2} \tag{8}$$

Putting

$$E = E(r, \psi) \cdot \exp[\text{i}(\omega t - \beta z)]$$

where $\omega = 2\pi f$ is the angular frequency of the wave and β is the propagation constant $(\beta = 2\pi f/\lambda)$ and inserting this expression into equation (8) gives

$$\Delta E(r, \psi) + (\omega^2 \mu \varepsilon - \beta) E(r, \psi) = 0 \tag{9}$$

The wave equation in cylindrical co-ordinates can be written:

$$\frac{\partial^2 E}{\partial r^2} + \frac{\partial E}{r \partial r} + \frac{1}{r^2} \cdot \frac{\partial^2 E}{\partial \psi^2} + (\omega^2 \varepsilon \mu - \beta^2) E = 0 \tag{10}$$

3.2. Field equations

The fields can be decomposed into longitudinal fields (parallel to z) $E_z(r, \psi)$ and transverse fields (orthogonal to z) $E_r(r, \psi)$:

$$E(r, \psi) = E_r(r, \psi) + E_z(r, \psi) \tag{11}$$

applying Maxwell's equations gives

$$E_r = p\left(\beta \, \frac{\partial E_z}{\partial r} + \frac{\omega \mu \partial H_z}{r \partial \psi}\right) \tag{12}$$

$$E_\psi = -p\left(-\frac{\beta \partial E_z}{r \partial \psi} + \omega \mu \, \frac{\partial H_z}{\partial r}\right) \tag{13}$$

$$H_r = -p\left(\frac{\omega \varepsilon \partial E_z}{r \partial \psi} - \beta \, \frac{\partial H_z}{\partial r}\right) \tag{14}$$

$$H_\psi = p\left(\omega \varepsilon \, \frac{\partial E_z}{\partial r} + \frac{\beta \partial H_z}{r \partial \psi}\right) \tag{15}$$

where p is a constant with a value given by

$$p = \frac{-i}{(\omega^2 \mu \varepsilon - \beta^2)}$$

4. ANALYTIC SOLUTION

To obtain an analytic solution, separation of variables is used, putting

$$E_z(r, \psi, t) = E_z(r) E_z(\psi) \cdot \exp[i(\omega t - \beta z)] \tag{16}$$

and inserting this into equation (10):

$$\frac{E_z''(r)}{E_z(r)} + \frac{E_z'(r)}{r E_z(r)} + \frac{E_z''(\psi)}{r^2 E_z(\psi)} + (\omega^2 \varepsilon \mu - \beta^2) = 0 \tag{17}$$

Putting

$$E_z(\psi) = C_1 \cdot \exp[-if\psi] + C_2 \cdot \exp[if\psi] \tag{18}$$

gives the coordinates of the field as a function of ψ in the z direction and f defines the azimuthal periodicity of the field and is an integer which defines the mode. The radial field equation can then be written

$$\frac{\partial E_z^2(r)}{\partial r^2} + \frac{1}{2} \cdot \frac{\partial E_z(r)}{\partial r} + (k^2 n^2(r) - \beta^2 - f^2/r^2)E_z(r) = 0 \tag{19}$$

All these calculations have used the E field but the same applies to the H field. The solution of equation (19) and the application of boundary conditions enable the fields propagating in the fibre to be determined analytically. This precise method is difficult to apply to GRIN fibres and will be applied only in the particular case of monomode optical fibres (see Chapter 3).

5. SOLUTION BY THE WBKJ METHOD

This method of calculation, associated with the refractive index profile, enables the modes excited in the fibre, their propagation constant and the associated modal dispersion to be determined. For this, the E wave function is expressed in the form

$$E(r) = E_0 \cdot \exp[ikS(r)] \tag{20}$$

where $S(r)$ is the function to be determined and is inserted into equation (19):

$$ikS'' - k^2 S'^2 + \frac{ik}{r} S' + k^2 \left(n^2(r) - \frac{\beta^2}{k^2} - \frac{f^2}{r^2 k^2} \right) = 0 \tag{21}$$

It is assumed that the variation of refractive index along a radial distance is small in comparison with the wavelength λ; a limited expansion with respect to k can then be made:

$$S(r) = S_0(r) + k^{-1}S_1(r) + k^{-2}S_2(r) + \cdots \tag{22}$$

This expression for $S(r)$ is inserted into equation (19) and it is assumed that the coefficients of k are independent of each other (the WBKJ approximation); hence

$$-k^2 \left[S_0'^2 - \left(n^2 - \frac{\beta^2}{k^2} - \frac{f^2}{k^2 r^2} \right) \right] = 0 \tag{23}$$

$$ikS_0'' - 2kS_0'S_1' + \frac{ik}{r} S_0' = 0 \tag{24}$$

solution of equations (23) and (24) gives

$$S_0(r) = \pm \frac{1}{k} \int_0^r \left(k^2 n^2(r) - \beta^2 - \frac{f^2}{r^2} \right)^{1/2} dr \qquad (25)$$

$$S_1(r) = \frac{i}{4} \cdot \log\left[r^2 n^2(r) - \frac{\beta^2 r^2}{k^2} - \frac{f^2}{k^2} \right] \qquad (26)$$

the first-order wave equation can then be written

$$E(r) = \frac{C_1 k}{rq^{1/2}} \cdot \exp\left[i \int_0^r q(r)\, dr \right] + \frac{C_2 k}{rq^{1/2}} \cdot \exp\left[-i \int_0^r q(r)\, dr \right] \qquad (27)$$

where

$$q^2 = k^2 n^2(r) - \beta^2 - \frac{f^2}{r^2} \qquad (28)$$

q determines the propagation constant which depends on r.

6. INTERPRETATION OF THE WAVE FUNCTION

6.1. The general wave equation

By combining the results of Sections 4 and 5, the field equation can be written

$$E_z(r, \psi, z, t) = (C_1 \exp[-if\psi] + C_2 \exp[if\psi]) \cdot \exp[i(\omega t - \beta z)]$$

$$\left(\frac{C_3 k}{rq^{1/2}} \exp\left[i \int_0^r q(r)\, dr \right] + \frac{C_4 k}{rq^{1/2}} \exp\left[-i \int_0^r q(r)\, dr \right] \right) \qquad (29)$$

Knowledge of the field E_z enables the radial and azimuthal fields to be determined. The factor β defines the projection of the wave vector in the z direction, f defines the fundamental frequency of the rotation ψ and q is a function of the radial distance. It is, therefore, this parameter which determines the various types of field which can propagate in the fibre and the evanescent ones which will be absorbed into the cladding. Differentiation of these fields is achieved by analysis of the function $q(r)$ which has two poles r_1 and r_2 for which it is zero. The boundary condition for $f \neq 0$ requires that $E(0) = 0$ and continuity of the fields through each area gives

$$C_1 = C_2 = 0$$

$r \leqslant r_1$

$$E(r) = \frac{Ck}{rq^{1/2}} \exp\left[- \int_r^{r_1} q(r)\, dr \right] \qquad (30)$$

$r_1 \leqslant r \leqslant r_2$

$$E(r) = \frac{Ck}{rq^{1/2}} \cos\left[\int_{r_1}^{r_2} q(r)\,dr - \frac{\pi}{4} \right] \qquad (31)$$

and for $r_2 \leqslant r \leqslant a$

$$E(r) = \frac{Ck}{rq^{1/2}} \left(\sin(\phi_1)\exp\left[-\int_{r_2}^{r} q(r)\,dr \right] + 2\cos(\phi_1)\exp\left(\int_{r_2}^{r} q(r)\,dr \right] \right) \qquad (32)$$

where

$$\phi_1 = \int_{r_1}^{r_2} q(r)\,dr$$

Neglecting the presence of the cladding ($r = \infty$), the boundary condition can be written as follows:

$$E(r \to \infty) = 0 \qquad \text{and so} \qquad \cos(\phi_1) = 0 \qquad (33)$$

$$E(r) = \left(\mu + \frac{1}{2} \right)\pi = \int_{r_1}^{r_2} q(r)\,dr$$

Since the number of modes is large, it is assumed tht $(\mu + 1)\pi \approx \mu\pi$. Thus every mode can be represented by two numbers, f for the azimuth and μ for the radial direction, and this implies discretization of the propagating fields.

6.2. Application

If the refractive index profile defined by equation (2) is used and the effective refractive index gradient is defined by

$$n_{\text{eff}} = \frac{\beta}{k} \qquad (34)$$

equation (28) can be rewritten by setting it to zero to obtain the roots r_1 and r_2:

$$n^2(r) = \frac{\beta^2}{k^2} + \frac{f^2}{r^2 k^2} \qquad (35)$$

By using the definition of the refractive index profile, the order of the mode is obtained:

$$n_1^2(1 - 2\Delta R^\infty) = n_{\text{eff}}^2 + \frac{f^2}{k^2} \cdot \frac{a^2}{R^2} \qquad (36)$$

Performing the calculation for a parabolic refractive index profile ($\alpha = 2$), the following roots are obtained ($r_1 = R_1/a$ and $r_2 = R_2/a$):

$$R_{1,2}^2 = \frac{1}{4n_1^2\Delta} \left((n_1^2 - n_{\text{eff}}^2) \pm \left[(n_1^2 - n_{\text{eff}}^2)^2 - \frac{8n_1^2\Delta a^2 f^2}{k^2} \right]^{1/2} \right) \qquad (37)$$

substituting into equation (35)

$$f = \frac{k}{a}\, n_1 (2\Delta)^{1/2} R_1 R_2 \qquad (38)$$

The $f = 0$ modes are given by

$$R_1 = 0 \qquad R_2 = \frac{n_1^2 - n_{\text{eff}}^2}{2\Delta n_1^2} \qquad (39)$$

which defines a category of modes which can propagate in the fibre.

7. THE INFLUENCE OF THE PROPAGATION CONSTANT

It should be observed that definition of the propagation constant $q(r)$ determines the type of wave which propagates in the fibre:

$$q^2(r) = k^2 n^2(r) - \beta^2 - \frac{f^2}{r^2} \qquad (40)$$

Several cases can occur:

$$\beta > kn_1$$
$$kn_1 \geqslant \beta \geqslant kn_2$$
$$kn_2 > \beta$$

where, in each case:

$$k^2 n^2 - \beta^2 > \frac{f^2}{r^2} \qquad \text{propagating mode}$$

$$k^2 n^2 - \beta^2 < \frac{f^2}{r^2} \qquad \text{evanescent mode}$$

7.1. Mode existence limits

Let $\beta > kn_1$: in this case

$$k^2 n_1^2 - \beta^2 < 0$$
$$k^2 n_2^2 - \beta^2 < 0$$

that is,

$$k^2 n^2 - \beta^2 < 0, \qquad \frac{f^2}{r^2} \geqslant 0 \tag{41}$$

In this configuration, the field is always evanescent; there is no guided propagation and the condition

$$\beta \leqslant k n_1 \tag{42}$$

is thus a propagation limiting condition.

7.2. The oscillating field

Consider the wave vectors such that $n_2 k \leqslant \beta \leqslant k n_1$: hence

$$k^2 n_1^2 - \beta^2 > 0$$
$$k^2 n_2^2 - \beta^2 < 0$$

in which case,

$$q^2(r) = k^2 n^2 - \beta^2 - \frac{f^2}{r^2} > 0 \tag{43}$$

that is

$$k^2 n^2 - \beta^2 > \frac{f^2}{r^2}$$

since $k^2 n^2 - \beta^2 > 0$, an oscillating field exists which can propagate in the fibre

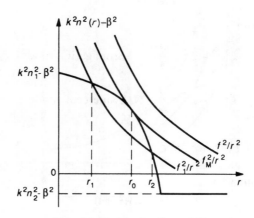

Figure 2.2   Region of oscillatory fields

if equation (43) is satisfied. Also

$$k^2 n_1^2 - \beta^2 > k^2 n_2^2 - \beta^2$$

A limiting value f_M exists for which equation (43) is not satisfied:

$$f_M^2 = r^2 (k^2 n^2 - \beta^2) \tag{44}$$

When $f > f_M$ all the fields are evanescent and there is no propagation. When $f < f_M$, a region exists in which the field is oscillatory: $r_1 < r < r_2$ and guided propagation occurs; outside this region, the fields are evanescent.

7.3. Leaky modes

The last case to be considered is that of $kn_2 \geqslant b$ where regions of evanescent and oscillatory fields can be defined:

$$k^2 n^2 - \beta^2 \geqslant \frac{f^2}{r^2}$$

If $f < f_m$, the modes can propagate for any value of r, but in the absence of confinement, their energy extends to infinity and they are rapidly absorbed by the difference of absorption coefficient between the core and the cladding. Another region exists where the modes are evanescent:

$$f_m < f < f_M$$

In conclusion, there are two regions where the evanescent fields can propagate; the first is situated in the core between r_{1c} and r_{2c} and the second beyond

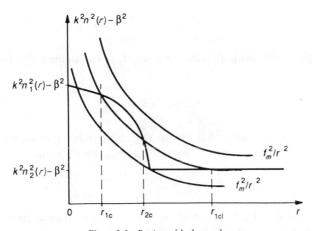

Figure 2.3 Region of leaky modes

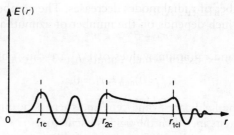

Figure 2.4   Radial field distribution

r_{1cl}. Part of the energy of these modes remains confined in the core of the fibre but these modes, described as leaky, contribute to the propagation only for a short distance.

8. THE NUMBER OF MODES IN A FIBRE

8.1. Calculation of the number of modes

Taking account of the degeneracy of each mode due to polarization states, the total number of modes propagating in an optical fibre can be written:

$$N = 4 \int_0^{f_M(\beta_0)} \mu \, df \qquad (45)$$

and by using equation (33):

$$N = \frac{4}{\pi} \int_0^{f_M(\beta_0)} \int_{r_1(f)}^{r_2(f)} q(r) \, dr \cdot df \qquad (46)$$

The single mode propagation constant $\beta_c = k n_2$ defines the largest radial mode:

$$\frac{f_M^2}{r^2} = k^2 n^2(r) - \beta_c^2 \qquad (47)$$

such that $\beta > \beta_c$ for any value of the refractive index parameter.

If equation (33) is rewritten, by replacing β_c by its value:

$$\mu\pi = \int_{r_1}^{r_2} \left(k^2 n^2(r) - k^2 n_2^2 - \frac{f^2}{r_1} \right)^{1/2} dr \qquad (48)$$

Observe that $\mu = 0$ for the value f_M which permits propagation and μ_{max} is minimal when $f = 0$; this indicates that as the number of azimuthal modes increases, the number of radial modes decreases. The number of radial modes has a maximum which depends on the number of azimuthal modes available in the cavity.

The maximum and minimum values of $r(f)$ are thus obtained for $f = 0$:

$$r(f = 0)_{min} = r_1 = 0$$
$$r_2(f = 0) = a$$
$$r(f = 0)_{max} = r_2(f = 0)$$

The maximum number of modes can be written

$$N = \frac{4}{\pi} \int_0^{r_2(f=0)} \left(k^2 n^2(r) - k^2 n_2^2 - \frac{f^2}{R} \right)^{1/2} dr \cdot df \qquad (49)$$

A change of variable can be made as follows:

$$x = f/r \qquad X = k^2 n^2(r) - \beta_c$$

This evaluates to

$$N = \int_0^{r_2(0)} k^2 n^2(r) r \, dr - \beta^2 \frac{r_2^2(0)}{2} \qquad (50)$$

If account is taken of the expression for the refractive index of the medium,

$$n(\beta) = \left[\frac{k^2 n_1^2 - \beta^2}{2} - \frac{2\Delta k^2 n_1^2}{\alpha + 2} \left(\frac{r_2(0)}{2} \right)^\alpha \right] r_2^2(0) \qquad (51)$$

Now

$$r_2^2(0) = a^2 \left(\frac{k^2 - \beta^2}{2\Delta k^2 n_1^2} \right)^{2/\alpha} \qquad (52)$$

By putting $\beta_c = kn_2$, the maximum number of modes up to cut-off is finally obtained:

$$N = \frac{\alpha}{\alpha + 2} a^2 k^2 n_1 \Delta \qquad (53)$$

The number of modes depends on the refractive index parameters, the wavelength, the core radius and the Δ of the refractive index. For a fibre optic of the parabolic type, this expression becomes

$$N_{\alpha=2} = \frac{a^2 k^2 n_1 \Delta}{2} \qquad (54)$$

As a numerical example, consider the case of a fibre with a core of radius 25 μm, $\Delta = 10^{-2}$ and refractive index $n_1 = 1.45$; the number of modes as a function of wavelength is as follows:

	$N(\lambda = 0.83)$	$N(\lambda = 1.3)$
$\alpha = 2$	376	153
$\alpha = \infty$	752	306

   In a fibre optic link, the number of modes propagating in a transmitting fibre must be identical to those propagating in a receiving fibre; otherwise mismatching occurs with a consequent loss of flux for which an expression can be derived from equation (53):

$$P_{dB} = 10 \cdot \log \frac{N_2}{N_1} = 10 \cdot \left[\log \frac{\Delta_2}{\Delta_1} + \log \frac{a_2^2}{a_1^2} + \log \frac{(\alpha_1 + 1)\alpha_2}{(\alpha_2 + 1)\alpha_1} \right] \qquad (55)$$

8.2. The relation between the number of modes and the propagation constant

It can be shown that, for a parabolic profile, the modes given by

$$m = 2\mu + |f|$$

have the same value of β; m is called the order of the group of modes and there are $m + 1$ modes which satisfy the degeneracy rule:

$$F = m, m - 2, m - 4, ..., 2, 0, -2, ..., -m$$

the number of modes can be written

$$n = \sum_1^m (m + 1) = \frac{1}{2}(m^2 + m) \approx \frac{m^2}{2}$$

hence for two polarization states the number of modes is $n = m^2$, and

$$M = n^{1/2} = akn_1 \left(\Delta \frac{\alpha + 2}{2} \right)^{1/2} \qquad (56)$$

Using equations (51) and (53), a number of modes n/N can be defined which is normalized with respect to the cut-off frequency:

$$\beta^2 = k^2 n_1^2 \left[1 - 2\Delta \left(\frac{n}{N} \right)^{\alpha/(\alpha+2)} \right] \qquad (57)$$

or, using standardized modes:

$$\beta = kn_1 \left[1 - 2\Delta \left(\frac{m}{M} \right)^{2\alpha/(\alpha+2)} \right]^{1/2} \qquad (58)$$

The separation between modes is

$$\Delta\beta m = \beta_m - \beta_{m-1} = \frac{d\beta}{dm} = -\frac{2\Delta^{1/2}}{a}\left(\frac{\alpha}{\alpha+2}\right)^{1/2}\left(\frac{m}{M}\right)^{\alpha-2/\alpha+2} \tag{59}$$

The distance between modes depends on the normalized order except in the case of a parabolic profile:

$$\Delta\beta_{\alpha=2} = -\frac{(2\Delta)^{1/2}}{a} \tag{60}$$

for a step index fibre it becomes:

$$\Delta\beta_{\alpha=\infty} = -\frac{2\Delta^{1/2}}{a}\cdot\frac{m}{M} = -\frac{2\Delta^{1/2}}{a}\cdot\left(\frac{n}{N}\right)^{1/2} \tag{61}$$

8.3. The influence of the injection conditions into the fibre

From the point of view of geometrical optics, the propagation constant can be written:

$$\vec{\beta_m} = kn(r)\cos\theta_m(r) \tag{62}$$

where θ_m is the injection angle of order m. Let θ_e be the angle of incidence of the light wave on the fibre (Fig. 2.5).
Snell's law gives:

$$\sin\theta_e = n(r)\cdot\sin\theta_i$$

that is

$$\beta^2 = k^2(n^2(r) - \sin^2(\theta_e)) \tag{63}$$

Combining equations (51) and (52), the relation between the orders which propagate and the injection conditions into the fibre is obtained

$$\frac{m}{M} = \left[\left(\frac{r}{a}\right)^\alpha + \frac{\sin^2\theta_e}{\sin^2\theta_c}\right]^{(\alpha+2)/2\alpha} \tag{64}$$

where $\sin\theta_e = n_1(2\Delta)^{1/2}$ is the numerical aperture of the fibre. If these results

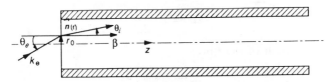

Figure 2.5 Injection into a fibre

are applied to step index fibres

$$\beta_m = kn_1 \cos \theta_m(r) \qquad (65)$$

that is

$$\frac{m}{M} = \frac{\sin \theta_e}{\sin \theta_c} = \frac{\theta_e}{\theta_c} \qquad (66)$$

and the separation between modes can be written:

$$\Delta\beta_{\alpha=\infty} = \frac{2\Delta^{1/2}}{a} \cdot \frac{\theta_e}{\theta_c} \qquad (67)$$

9. DISPERSION IN OPTICAL FIBRES

Dispersion in fibres is a combination of three types:

— Intramodal dispersion,
— Intermodal dispersion,
— Chromatic dispersion,

The disadvantage of dispersion is spreading and distortion of the luminous pulse carried by the fibre; this reduces the bandwidth.

9.1. Intramodal dispersion

This dispersion is due to the dispersive properties of the material and the waveguide structure. For a given mode, the variation of refractive index with wavelength causes a spreading of the signal which can be large in monomode fibres but negligible in multimode fibres.

9.2. Intermodal dispersion

Intermodal dispersion is associated with the refractive index profile and is obtained from the definition of the group delay time for a given wavenumber:

$$\tau(\beta) = L \cdot \frac{d\beta}{d\omega} \qquad (68)$$

where L is the length of the fibre.

$$\tau(\beta) = L \cdot \frac{d\beta}{dk_1} \cdot \frac{dk_1}{d\lambda} \cdot \frac{d\lambda}{d\omega} \qquad (69)$$

and $k_1 = kn_1$; putting the group refractive index as

$$N_1 = n_1 - \frac{dn_1}{d\lambda} \cdot \lambda \tag{70}$$

$$\tau(\beta) = L \cdot \frac{N_1}{c} \cdot \frac{d\beta}{dk_1} \tag{71}$$

equation (58) gives an expression for β as a function of the order of the mode:

$$\beta = kn_1 \left[1 - 2\Delta \left(\frac{m}{M} \right)^{2\alpha(\alpha+2)} \right]^{1/2}$$

replacing m and M from equation (56) and differentiating:

$$\frac{d\beta}{dk} = \frac{k_1}{\beta} - \left(\frac{\Delta k_1^2}{\beta} \right)^{\alpha/(\alpha+2)} \cdot \frac{k_1^2 - \beta^2}{2\Delta k_1^2}$$

$$\times \left[\frac{4}{\alpha+2} \Delta^{2/\alpha+2} \cdot k_1^{(2-\alpha)/(\alpha+2)} + \frac{2}{\alpha+2} k_1^{4/(\alpha+2)} \Delta^{-\alpha/(\alpha+2)} \frac{d\Delta}{dk_1} \right] \tag{72}$$

The expression $d\beta/dk_1$ represents the dispersion of the refractive index profile as a function of wavelength:

$$\frac{d\Delta}{dk_1} = \frac{d\Delta}{d\lambda} \cdot \frac{d\lambda}{dk_1} = -\frac{\lambda}{k_1} \cdot \frac{d\Delta}{d\lambda} \cdot \frac{n_1}{N_1} = -p \frac{\Delta}{k_1} \tag{73}$$

where p is the normalized dispersion of the refractive index profile. The mode dispersion equation can be written:

$$\frac{d\beta}{dk_1} = \frac{k_1}{\beta} \left[1 - \left(1 - \frac{\beta^2}{k^2 n_1^2} \right) \left(\frac{2-p}{\alpha+2} \right) \right] \tag{74}$$

and the equation for the dispersion of the group delay time is:

$$\tau(\beta) = \frac{LN_1}{c} \cdot \frac{kn_1}{\beta} \left[1 - \frac{2-p}{\alpha+2} \left(1 - \frac{\beta^2}{k^2 n_1^2} \right) \right] \tag{75}$$

The propagation constant β of the modes is subjected to dispersion, as is the group delay time, but each mode is not subjected in the same way and does not propagate in the same manner or at the same velocity in the fibre. Performing a limited expansion of this equation with respect to Δ:

$$\frac{kn_1}{\beta} = 1 + \Delta \left(\frac{m}{M} \right)^{2\alpha/(\alpha+2)} + \frac{3}{2} \Delta^2 \left(\frac{m}{M} \right)^{4\alpha/(\alpha+2)} + \cdots \tag{76}$$

hence, neglecting third order terms and putting $X = m/M$,

$$\tau(X) = \frac{LN_1}{c} \left[1 - \Delta X^{2\alpha/(\alpha+2)} \left(\frac{2 - 2p - \alpha}{\alpha + 2} \right) - \Delta^2 X^{4\alpha/(\alpha+2)} \left(\frac{2 + 4p - 3\alpha}{2\alpha + 4} \right) \right]$$

(77)

The separation between the lowest and highest modes can be written

$\Delta\tau = \tau(X) - \tau(0)$

$$= \frac{LN_1}{c} \left[\Delta X^{2\alpha/(\alpha+2)} \left(\frac{\alpha - 2 + 2p}{\alpha + 2} \right) + \Delta^2 X^{4\alpha/(\alpha+2)} \left(\frac{3\alpha + 4p - 2}{2\alpha + 4} \right) \right]$$

(78)

This difference of propagation between modes limits the bandwidth of the fibre since, for two distinct pulses of light, the slowest modes of the first pulse must not become mixed with the fastest of the second.

9.3. Chromatic dispersion

Silica is a dispersive optical medium for which the dependance can be established by using a limited expansion of the Sellnier type:

$$n^2 - 1 = \sum_{i=1}^{3} \frac{A_i \lambda^2}{\lambda^2 - B_i^2}$$

(79)

This relationship corresponds to the influence of the three principal absorption bands, each of which is distinguished by the constants A_i and B_i. Hence, for a source of spectral density $S(\lambda)$ centred on λ_0,

$$\tau(\lambda, X) = \tau(\lambda_0, X) + (\lambda - \lambda_0) \left. \frac{d\tau(\lambda, X)}{d\lambda} \right|_{\lambda_0}$$

(80)

to a first order; the first term corresponds to the intermodal dispersion and the second to the chromatic dispersion. The chromatic dispersion parameter is defined by

$$M_d = \left. \frac{d\tau(\lambda, X)}{d\lambda} \right|_{\lambda_0} = - \frac{\lambda_0}{c} \cdot L \cdot \left. \frac{d^2 n}{d\lambda^2} \right|_{\lambda_0}$$

(81)

which can be written:

$$M_d = - \frac{\lambda_0}{c} \cdot \frac{L}{n} \left[4\lambda^2 \sum_{i=1}^{3} \frac{A_i B_i^2}{(\lambda^2 - B_i^2)^3} + \frac{n}{\lambda} \cdot \frac{dn}{dL} - \left(\frac{dn}{d\lambda} \right)^2 \right]$$

(82)

The coefficient M_d is zero for $\lambda = 1.278$ μm and is expressed in ps/nm · km; this explains the window at $\lambda = 1.27$ μm used in fibre optic telecommunication.

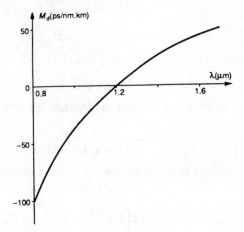

Figure 2.6 Variation of M_d as a function of wavelength

10. DETERMINATION OF THE OPTIMUM REFRACTIVE INDEX PROFILE

The optimum refractive index profile is that which provides the minimum dispersion for both intermodal dispersion and chromatic dispersion. The case of intramodal dispersion is examined in the context of monomode optical fibres where its role is predominant.

10.1. The optimum refractive index profile for intermodal dispersion

This is distinguished by the absence of propagation differences between the modes:

$$\frac{d\Delta\tau}{dX} = 0 = \Delta \frac{2\alpha}{\alpha+2} X^{(\alpha-2)(\alpha+2)}[(\alpha-2+2p) + \Delta(3\alpha+4p-2)X^{2\alpha/(\alpha+2)}]$$

$$(83)$$

for $X = 0$:

$$\frac{d\Delta\tau}{dX} = \alpha - 2 + 2p = 0$$

$$\alpha = 2 - 2p$$

for $X = 1$:

$$\alpha = 2 - 2p - 4\Delta + 2p\Delta$$

To eliminate intermodal dispersion it is sufficient that

$$2 - 2p - 4\Delta + 2p\Delta < \alpha < 2 - 2p \qquad (84)$$

There is another condition for which intermodal dispersion is zero:

$$X^{2\alpha/(\alpha+2)} = \frac{2 - \alpha - 2p}{\Delta(3\alpha + 4p - 2)} \qquad (85)$$

This corresponds to the maximum number of modes; the associated value of X is obtained for

$$\alpha_{\text{opt}} = 2 - 2p - 2\Delta\left(1 - \frac{p}{2}\right) \approx 2 - 2p - 2\Delta \qquad (86)$$

The modes corresponding to $X = 0$ and $X = 1$ arrive at the same time; if $\alpha > \alpha_{\text{opt}}$ the high-order modes arrive after the low-order modes and for $\alpha < \alpha_{\text{opt}}$ the low-order modes arrive after the high-order modes.

Consider the case of a parabolic profile, the maximum dispersion between high- and low-order modes is

$$\Delta\tau(\alpha = 2) = \frac{LN_1}{c}\left[\frac{\Delta p}{2} + \frac{\Delta^2}{2}(1 + p)\right] \qquad (87)$$

In comparison with a step index fibre, a factor of 800 in index profile can be gained; moreover it can be seen (Figure 2.7) that a very small change in the optimum value has a significant effect on the separation between modes.

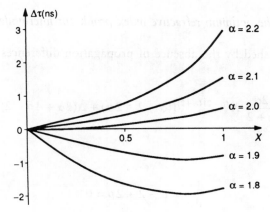

Figure 2.7 Variation of propagation time with normalized mode

10.2. The optimum index profile for chromatic dispersion

The normalized refractive index dispersion parameter p can be written

$$p = \frac{n_1}{N_1} \cdot \frac{\lambda}{\Delta} \cdot \frac{d\Delta}{d\lambda} \tag{88}$$

It is a parameter which depends on the dopants used in producing the optical fibre and the compound $P_2O_6 - SiO_2$ has the advantage of a small variation of the parameter p as a function of wavelength; the refractive index parameter α_{opt} also varies little with wavelength. In general, the aim is to optimize fibres for a single wavelength; this is more easily achieved than for a wide spectrum.

11. VARIATION OF REFRACTIVE INDEX PROFILE

11.1. Intrinsic variation of the refractive index profile

The refractive index profile plays an important part in determining the spread in propagation times between modes ($X = 1$)

$$\Delta\tau = \frac{LN_1}{c} \Delta \left[\left(\frac{\alpha - 2 + 2p}{\alpha + 2} \right) + \Delta \frac{3\alpha + 4p - 2}{2\alpha + 4} \right] \tag{89}$$

This variation is thus minimized for a value of refractive index parameter slightly less than 2; between $\alpha = 1.9$ or 2.1 and $\alpha = 2$ there is a reduction by a factor of 5 in the spread in propagation times.

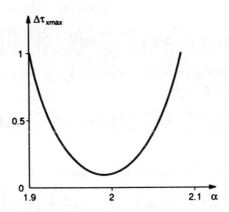

Figure 2.8   Variation of $\Delta\tau$ as a function of refractive index parameter

11.2. Longitudinal variation in the refractive index profile

The refractive index parameter is not constant along the fibre; it is subject to variation. Two simple analytical approaches can be considered; the first represents linear variation of the index parameter and the second a periodic variation.

11.2.1. Linear variation of the refractive index profile

It is assumed that the refractive index parameter varies linearly between the two ends of the fibre:

$$\alpha(z) = \alpha_1 + \frac{\alpha_1 - \alpha_2}{L} z \tag{90}$$

where α_1 and α_2 are the refractive index parameters at the ends of the fibre. In this case, the mean value of the dispersion can be written:

$$\langle \Delta \tau_x \rangle = \frac{1}{L} \int_0^L \Delta \tau(X, a) \, dz \tag{91}$$

On replacing $\Delta \tau(X, \alpha)$ by the value of equation (77), this becomes

$$\langle \Delta_{\tau_x} \rangle = \frac{N_1}{c} \Delta X^{2\alpha/(\alpha+2)} \int_0^L \left[\frac{\alpha(z) - 2 + 2p}{\alpha(z) + 2} + \Delta x^2 \frac{3\alpha(z) + 4p - 2}{2\alpha(z) + 4} \right] dz \tag{92}$$

assuming that $2\alpha/(\alpha + 2)$ is approximately constant along the fibre and integrating one obtains

$$\langle \Delta_{\tau_x} \rangle = \frac{LN_1}{c} \Delta X^{2\alpha/(\alpha+2)} \left\{ \left[\frac{-4 + 2p}{\alpha_2 - \alpha_1} \ln\left(\frac{\alpha_2 + 2}{\alpha_1 + 2} \right) \right] \right.$$
$$\left. + 1 + \Delta X^2 \left[\frac{2p - 4}{\alpha_2 - \alpha_1} \ln\left(\frac{\alpha_2 + 2}{\alpha_1 + 2} \right) + \frac{3}{2} \right] \right\} \tag{93}$$

Taking the case of the extreme modes $X = 1$ and assuming the normalized dispersion parameter p is zero,

$$\langle \Delta \tau \rangle = \frac{LN_1}{c} \Delta \left[\frac{-4}{\alpha_2 - \alpha_1} \ln\left(\frac{\alpha_2 + 2}{\alpha_1 + 2} \right) + 1 - \Delta \frac{4}{\alpha_2 - \alpha_1} \ln\left(\frac{\alpha_2 + 2}{\alpha_1 + 2} \right) + \frac{3}{2} \right] \tag{94}$$

An expansion limited to the third order gives

$$\langle \Delta \tau \rangle = \frac{LN_1}{c} \Delta \left[\frac{-4}{\alpha_2 - \alpha_1} \ln\left(\frac{\alpha_2 + 2}{\alpha_1 + 2} \right) + 1 \right] \tag{95}$$

which permits the value of the mean refractive index parameter to be defined:

$$\langle \alpha \rangle = \frac{\alpha_2 - \alpha_1}{\ln\left[(\alpha_2 + 2)/(\alpha_1 + 2)\right]} - 2 \tag{96}$$

In the case where $\alpha_1 = 2$ and $\alpha_2 = 2.1$ one obtains $\langle \alpha \rangle = 2.0498$, which is slightly greater than the optimum value of the index profile; there will therefore be intermodal dispersion. This variation is particularly prominent on the core boundary (Fig. 2.9).

11.2.2. Periodic variation of the refractive index profile

A periodic variation of the index profile will now be assumed such that

$$\alpha(z) = \alpha_1 + \delta\alpha \, \sin\left(\frac{2\pi N_0}{L} z\right) \tag{97}$$

The mean variation of the intermodal dispersion to first order, can be written

$$\langle \Delta \tau_x \rangle = \frac{LN_1}{c} \, \Delta X^{2\alpha/(\alpha+2)} \left[1 + \frac{2p-4}{\pi N_0 D} \, W + \Delta X^2 \left(\frac{3}{2} - \frac{2p-4}{\pi N_0 D} \, W\right)\right] \tag{98}$$

with:

$$W = \arctan\left[\frac{(\alpha_1 + 2)\tan(\pi N_0) + \delta\alpha}{D}\right] - \arctan\left[\frac{\delta\alpha}{D}\right]$$

$$D = \left[(\alpha_1 + \alpha_2)^2 - \delta\alpha^2\right]^{1/2} \tag{99}$$

which can be rewritten for the case of higher-order modes for $X = 1$, with the

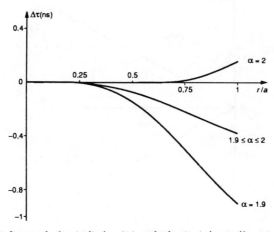

Figure 2.9 The influence of a longitudinal variation of refractive index profile on intermodal dispersion

assumption that $p = 0$ and to the third order:

$$\langle \Delta \tau \rangle = \frac{L N_1}{c} \Delta \left[1 - \frac{4}{\pi N_0 D} W \right] \tag{100}$$

It can be seen from Figure 2.10 that a variation $\delta \alpha = 0.1$ of the index parameter causes a change of 80% with respect to the undisturbed value.

11.3. Radial variation of the refractive index profile

Use of the vapour phase deposition process allows a refractive index profile to be realized which is approximated by a series of steps of varying size according to the deposits, and this leads to a variation of the index parameter:

$$n^2(r) = \begin{cases} n_{10}^2 \left[1 - \Delta_1 \left(\dfrac{r}{a_1} \right)^{\alpha_1} \right] & 0 \leqslant r \leqslant b \\[2em] n_{20}^2 \left[1 - \Delta_2 \left(\dfrac{r}{a_2} \right)^{\alpha_2} \right] & b \leqslant r \leqslant a_2 \end{cases} \tag{101}$$

The boundary conditions are observed together with

$$n_{10}^2 \Delta_1 \left[1 - \left(\frac{b}{a_1} \right)^{\alpha_1} \right] = n_{20}^2 \Delta_2 \left[1 - \left(\frac{b}{a_2} \right)^{\alpha_2} \right]$$

Taking account of the 'step' in the index parameter, the expression for the

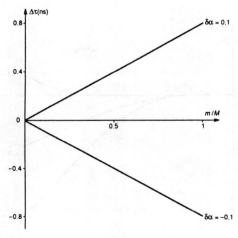

Figure 2.10   The influence of periodic and longitudinal variation of the refractive index parameter

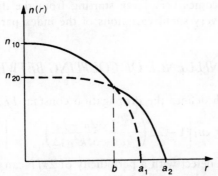

Figure 2.11 Refractive index profile with a double valued index parameter

number of modes is

$$n(\beta) = (k^2 n_{10}^2 - k^2 n_{20}^2)\,\frac{b^2}{2} + a_2^2 \Delta_2 k^2 n_{20}^2\,\frac{\alpha_2}{\alpha_2 + 2}\left(\frac{k^2 n_{20}^2 - \beta^2}{2\Delta_2 k^2 n_{20}^2}\right)^{(\alpha_2 + 2)/\alpha_2}$$

$$-2\left[\frac{\Delta_1 k^2 n_{10}^2}{\alpha_1 + 2}\left(\frac{b}{a_1}\right)^{\alpha_1} - \frac{\Delta_2 k^2 n_{20}^2}{\alpha_2 + 2}\left(\frac{b}{a_2}\right)^{\alpha_2}\right]b^2 \quad (102)$$

the group delay time can be deduced:

$$\tau(\beta) = \frac{LN_1}{c}\left(\left[\frac{\beta}{kn_{10}} + (2\Delta_2)\right]^{(\alpha_2 + 2)/\alpha_2}\frac{kn_{10}}{\beta}\right]\cdot\left[1 - \frac{\beta^2}{k^2 n_2^2}\right]^{-2/\alpha_2}$$

$$\times\left[\frac{n_{20}^2}{n_{10}^2}\cdot\frac{\alpha_2}{\alpha_2 + 2}\left(\frac{1 - \beta^2/k^2 n_{20}^2}{2\Delta_2}\right)^{(\alpha_2 + 2)/\alpha_2} - \frac{\Delta_1}{\Delta^2}\cdot\frac{2}{\alpha_1 + 2}\cdot\frac{b^{\alpha_1}}{a_1^{\alpha_1}}\cdot\frac{b^2}{a_2^2}\right.$$

$$\left.+ \frac{n_{20}^2}{n_{10}^2}\cdot\frac{2}{\alpha_2 + 2}\left(\frac{b}{a_2}\right)^{\alpha_2 + 1} + \left(1 - \frac{n_{20}^2}{n_{10}^2}\right)\frac{b^2}{a_2^2}\cdot\frac{1}{2\Delta_2}\right] - 1\right) \quad (103)$$

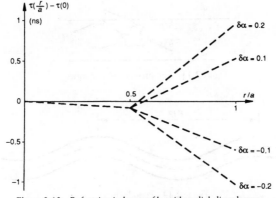

Figure 2.12 Refractive index profile with radial disturbance

The disperion becomes very large starting from the discontinuity in the parameter even for very small variations of the index parameter.

12. THE INFLUENCE OF COUPLING BETWEEN MODES

Equation (58) which defines the propagation constant $b_{f,\mu}$ can be written:

$$\beta_{f,\mu} = kn_1\left[1 - 2\Delta\left(\frac{\alpha + 2}{\alpha} \cdot \frac{(2\mu + f)^2}{a^2k^2n_1^2\Delta}\right)^{\alpha/(\alpha + 2)}\right]^{1/2} \tag{104}$$

The presence of a defect with a periodicity of $2\pi/F$ can couple two modes together on condition that

$$\Delta\beta_{f,\mu} = \beta_{\mu,f+1} - \beta_{\mu,f} = 2\pi/F \tag{105}$$

The main causes of coupling phenomena are the presence of impurities, periodic curvature and periodic constriction of the fibre. In the case of a fibre with a parabolic index profile, this condition can be written

$$\Delta\beta_{f,f+1} = \frac{(2\Delta)^{1/2}}{a} = \frac{2\pi}{F} \tag{106}$$

let

$$F = a\pi(2/\Delta)^{1/2} \tag{107}$$

For a step index fibre,

$$F = \frac{1}{a^2kn_1}(1 + 2(2\mu + f)) \tag{108}$$

The separation between modes is a function of the order of the mode;

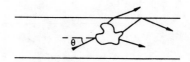

Figure 2.13   The presence of impurities

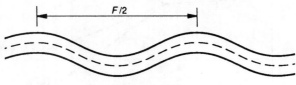

Figure 2.14   Periodic bending

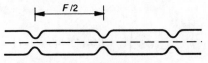

Figure 2.15 Periodic constriction

hence, taking account of the bounds of the guided modes:

$$\frac{1}{a^2 k n_1} \leqslant \Delta\beta \leqslant \frac{1}{a^2 k n_1} + \frac{2(\Delta)^{1/2}}{a} \tag{109}$$

Now consider a defect of period $2\pi/F$ of the optical fibre:

● There is no coupling if

$$\frac{2\pi}{F} < \frac{1}{a^2 k n_1}$$

● There is coupling between guided modes and radiation if

$$\frac{2\pi}{F} \geqslant \frac{1}{a^2 k n_1} + \frac{2(\Delta)^{1/2}}{a}$$

● There is coupling between guided modes and propagation if

$$\frac{1}{a^2 k n_1} \leqslant \frac{2\pi}{F} \leqslant \frac{1}{a^2 k n_1} + \frac{2(\Delta)^{1/2}}{a}$$

From these relations it is possible to realize optical couplers which enable part of the signal to be extracted. This principle has also been used for the realization of detectors.

13. THE FREQUENCY RESPONSE OF THE FIBRE

The impulse response of a uniformly excited fibre is described by

$$P(t) = \frac{1}{N} \cdot \frac{dn}{dt} = \frac{2m}{M^2} \cdot \frac{dm}{dt} = 2X \cdot \frac{dX}{dt} \tag{110}$$

Let $f(t)$ be the incident signal and $I(X)$ the injection condition; the response at a distance z can be written

$$P(z, t) = \int_0^1 2X \cdot I(X) \cdot f(t - \tau(X)z) \, dX \tag{111}$$

The frequency response is obtained by taking the Fourier transform of this function:

$$\Gamma(\omega, z) = \frac{1}{(2\pi)^{1/2}} \int_{-\infty}^{\infty} P(z, t) \exp(-i\omega t) \, dt \tag{112}$$

The transfer function of the optical fibre $H(\omega, z)$ is written

$$H(\omega, z) = \frac{\Gamma(\omega, z)}{\Gamma(\omega, 0)} \tag{113}$$

The modulus of the transfer function (MTF), assuming a Gaussian injection profile, can be written

$$I(X) = I_0 \exp\left(-\frac{X^2}{X_0^2}\right) \tag{114}$$

$$|H(\omega, z)| = 20 \log\left[X_0^2\left[1 - \exp\left(-\frac{1}{X_0^2}\right)\right](A^2 + B^2)^{1/2}\right] \tag{115}$$

with

$$A = \int_0^1 2X \exp\left(-\frac{X^2}{X_0^2}\right)\cos(\Omega\Delta\tau(X)) \, dX$$

$$B = \int_0^1 2X \exp\left(-\frac{X^2}{X_0^2}\right)\sin(\Omega\Delta\tau(X)) \, dX$$

$$\Omega = \frac{\omega n_1}{c} \Delta Z$$

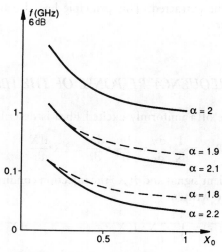

Figure 2.16   The -6 dB frequency in GHz as a function of X

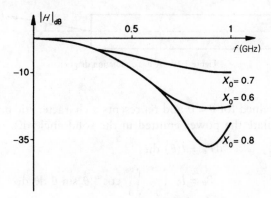

Figure 2.17   Transfer function as a function of frequency

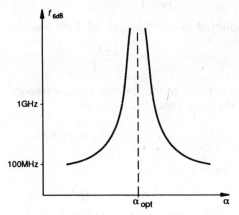

Figure 2.18   The −6 dB frequency as a function of the refractive index parameter

The variation in frequency for an MTF = −6 dB is represented in Figure 2.16 as a function of the various modes.

The MTF increases as X_0 decreases from which can be seen the advantage of a mode order $X \leqslant 0.4$.

If there were no dispersion, the frequency would be infinite for α_{opt}.

14. THE EFFECTIVE NUMERICAL APERTURE

To calculate the effective numerical aperture of the fibre, assume that the radiation pattern of the fibre can be defined by a law of the form:

$$I(\theta) = I_0 \cos^m(\theta) \tag{116}$$

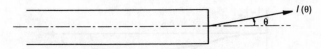

Figure 2.19 A basic radiation diagram

where I_0 is obtained for $\theta = 0$ and represents a characteristic parameter of the fibre. To calculate the power emitted in the solid angle $d\Omega$, one writes

$$dP_\theta = I(\theta)\ d\Omega \tag{117}$$

$$P_\theta = I_0 \int_0^{2\pi} \int_0^{\theta_0} \cos^m(\theta)\sin\theta\ d\theta\ d\psi \tag{118}$$

$$P_\theta = \frac{2\pi I_0}{m+1}\ [1 - \cos^{m+1}(\theta_0)] \tag{119}$$

The total power emitted in a solid angle of $2\pi(\theta = \pi/2)$ is

$$P_t = \frac{2\pi I_0}{m+1} \tag{120}$$

By definition, the effective numerical aperture corresponds to the solid angle in which 90% of the total power is emitted:

$$\frac{P_{\theta_0}}{P_t} = 0.9 \qquad \cos(\theta_0) = 0.1^{1/m+1} \tag{121}$$

Since in general, $m \gg 1$, one has

$$I(\theta_0) = 0.1 \cdot I_0 \tag{122}$$

The actual numerical aperture differs slightly from the theoretical one due to the limitation of geometrical optics in respect of real phenomena.

15. EXAMPLES OF MULTIMODE FIBRES

Table 2.1 presents the optical characteristics of three types of currently used fibres. The plastic fibre is the FORT SP 100, the step index fibre is the CLTO FGA 7000 and the graded index fibre is the FORT LG5040300.

The choice of fibre type depends on the application; it is evident that graded index optical fibres of optimum performance are used for high-capacity (100 MHz) and long-distance (several kilometres) communication. Step index fibres are used for short distances (of the order of a kilometre) and medium capacities; plastic fibres are used for very short distances (of the order of a hundred metres) and very low capacities.

Table 2.1 Comparison of three types of fibre (λ = 860 nm)

	Plastic (SF 100)	CLTO (FGA 7000)	FORT LG5040300
Material	polymer	silica	silica
Attenuation (dB/km)	500	7	3.5
Core diameter (μm)	980	100	50
Cladding diameter (μm)	1000	140	125
Bandwidth (MHz/km)	—	20	400
Numerical aperture	0.50	0.30	0.19
Fibre type	Step index	Step index	Graded index

3

MONOMODE OPTICAL FIBRES

1. INTRODUCTION

Monomode optical fibres have two major advantages; these are a wide bandwidth and low attenuation. The fact that a single mode is propagating limits chromatic dispersion which is caused by variation of refractive index as a function of wavelength. It is thus possible to select a spectral window which allows a frequency band greater than 10 GHz/km to be obtained. Propagation of a single mode also limits attenuation as a function of wavelength and this allows the distance between repeaters to be increased. By varying the optoelectronic parameters of the dielectric guides which constitute the fibre, it is possible to optimize them for a given wavelength; but this generally leads to a very small core diameter which causes coupling problems.

2. DEFINITION AND SOLUTION OF THE SCALAR WAVE EQUATION

In this chapter an exact solution of these equations will be obtained in order to deduce the main parameters of monomode optical fibres from both the fabrication and application points of view. It was shown in Chapter 2, on multimode optical fibres, that the transverse components of the electromagnetic E and H fields associated with the modes which propagate in the fibre can be expressed as a function of only the longitudinal components E_z and H_z. These components, in cylindrical coordinates, satisfy the equation

$$\frac{\partial^2 \psi}{\partial r^2} + \frac{1}{r} \cdot \frac{\partial \psi}{\partial r} + \frac{1}{r^2} \cdot \frac{\partial^2 \psi}{\partial \phi^2} + \chi^2(r)\psi = 0 \tag{1}$$

where ψ represents either E_z or H_z and χ^2 represents the propagation constant given by

$$\chi^2(r) = k^2 n^2 - \beta^2 \tag{2}$$

Using the method of separation of variables, one can write

$$\psi(r, \phi, z, t) = F(r) \cdot G(\phi) \cdot \exp[i(\omega t - \beta z)] \tag{3}$$

Equation (1) can be rewritten:

$$\frac{r^2}{F(r)} \cdot \left[\frac{d^2 F(r)}{dr^2} + \frac{1}{r} \cdot \frac{dF(r)}{dr} \right] + \chi^2 r^2 = -\frac{1}{G(\phi)} \cdot \frac{d^2 G(\phi)}{d\theta^2} \tag{4}$$

Each term of Equation (4) must be equal to a constant, from which

$$G(\phi) = G_1 \cdot \exp(\beta_\infty \phi) + G_2 \cdot \exp(-\beta_\infty \phi) \tag{5}$$

since:

$$G(\phi) = G(\phi + 2\pi)$$

one deduces that:

$$\beta_\infty^2 = -f^2 \quad \text{where} \quad f = 1, 2, 3, \ldots$$

and the form of $G(\phi)$ is

$$G(\phi) = G_1 \cdot \exp(-if\phi) + G_2 \cdot \exp(if\phi) \tag{6}$$

This function describes the variation of the fields E_z and H_z as a function of the angle of propagation ϕ in the fibre. If this solution is inserted into Equation (4) one obtains

$$\frac{d^2 F(r)}{dr^2} + \frac{1}{r} \cdot \frac{dF(r)}{dr} + \left[\chi^2 - \frac{f^2}{r^2} \right] \cdot F(r) = 0 \tag{7}$$

The general solution of this equation is a Bessel or Neumann function which, according to the sign of χ^2, can be written

$$F(r) = A \cdot J_f(\chi r) + B \cdot N_f(\chi r) \quad \chi^2 > 0 \tag{8}$$

where J_f is a Bessel function of the first kind of order f and N_f is a Neumann function of order f.

$$F(r) = C \cdot I_f(\gamma r) + D \cdot K_f(\gamma r) \quad \chi^2 < 0 \tag{9}$$

where $\gamma = -i\chi$, I_f is a modified Bessel function of order f and K_f is a McDonald function of order f.

3. APPLICATION TO A STEP INDEX MONOMODE OPTICAL FIBRE

Consider a step index fibre (Figure 3.1) of core radius a, cladding radius b, core refractive index n_1 and cladding refractive index n_2. It will be assumed here that b tends to infinity for the boundary conditions.

3.1. The field equations

If account is taken of the boundary conditions, one must have

- $r = \infty$ for zero fields
- $r = 0$ for the tangential components and their derivatives.

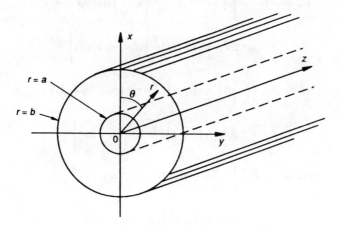

Figure 3.1 Fibre geometry

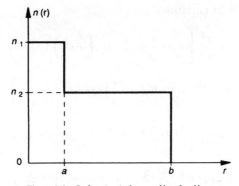

Figure 3.2 Refractive index profile of a fibre

Inserting these conditions into the equations of Section 2 and putting

$$
\begin{aligned}
u^2 &= a^2(k^2 n_1^2 - \beta^2) \qquad r \leqslant a \\
\mathbf{w}^2 &= a^2(\beta^2 - k^2 n_2^2) \qquad r \geqslant a
\end{aligned}
\tag{10}
$$

one obtains the equations for the various fields:

For $r \leqslant a$:

$$
\begin{aligned}
E_r = -\mathrm{i}\,\frac{a^2}{u^2}\,\bigg[&\beta \cdot \frac{u}{a} \cdot A_f^{\mathrm{I}} \cdot J_f'\left(u \cdot \frac{r}{a}\right)\cos(f\phi + \alpha_f) \\
&- \omega \cdot \frac{\mu_0}{r} \cdot B_f^{\mathrm{I}} \cdot J_f\left(u \cdot \frac{r}{a}\right)\cdot f \cdot \sin(f\phi + \psi_f)]
\end{aligned}
\tag{11}
$$

$$
\begin{aligned}
E_\phi = -\mathrm{i}\,\frac{a^2}{u^2}\,\bigg[&-\frac{\beta}{r} \cdot f \cdot A_f^{\mathrm{I}} \cdot J_f\left(u \cdot \frac{r}{a}\right)\sin(f\phi + \alpha_f) \\
&- \omega\mu_0 \cdot B_f^{\mathrm{I}} \cdot \frac{u}{a} \cdot J_f'\left(u \cdot \frac{r}{a}\right)\cos(f\phi + \psi_f)\bigg]
\end{aligned}
\tag{12}
$$

$$
\begin{aligned}
H_r = -\mathrm{i}\,\frac{a^2}{u^2}\,\bigg[&\beta \cdot \frac{u}{a} \cdot B_f^{\mathrm{I}} \cdot J_f'\left(u \cdot \frac{r}{a}\right)\cos(f\phi + \psi_f) \\
&+ \frac{\omega\varepsilon_1}{r} \cdot A_f^{\mathrm{I}} \cdot J_f\left(u \cdot \frac{r}{a}\right)\cdot f \cdot \sin(f\phi + \alpha_f)\bigg]
\end{aligned}
\tag{13}
$$

$$
\begin{aligned}
H_\theta = -\mathrm{i}\,\frac{a^2}{u^2}\,\bigg[&-\frac{\beta}{r} \cdot f \cdot B_f^{\mathrm{I}} \cdot J_f\left(u \cdot \frac{r}{a}\right)\sin(f\phi + \psi_f) \\
&+ \omega\varepsilon_1 \cdot A_f^{\mathrm{I}} \cdot J_f'\left(u \cdot \frac{r}{a}\right)\cos(f\phi + \alpha_f)\bigg]
\end{aligned}
\tag{14}
$$

For $r \geqslant a$, the field equations are

$$
\begin{aligned}
E_r = -\mathrm{i}\,\frac{a^2}{\mathbf{w}^2}\,\bigg[&-\beta \cdot \frac{w}{a} \cdot A_f^{\mathrm{II}} \cdot K_f'\left(w \cdot \frac{r}{a}\right)\cos(f\phi + \alpha_f) \\
&+ \omega \cdot \frac{\mu_0}{r} \cdot B_f^{\mathrm{II}} \cdot K_f\left(w \cdot \frac{r}{a}\right)\cdot f \cdot \sin(f\phi + \psi_f)\bigg]
\end{aligned}
\tag{15}
$$

$$
\begin{aligned}
E_\phi = -\mathrm{i}\,\frac{a^2}{\mathbf{w}^2}\,\bigg[&\frac{\beta}{r} \cdot f \cdot A_f^{\mathrm{II}} \cdot K_f\left(w \cdot \frac{r}{a}\right)\sin(f\phi + \alpha_f) \\
&+ \omega\mu_0 \cdot B_f^{\mathrm{II}} \cdot \frac{w}{a} \cdot K_f'\left(w \cdot \frac{r}{a}\right)\cos(f\phi + \psi_f)\bigg]
\end{aligned}
\tag{16}
$$

$$H_r = -\,i\,\frac{a^2}{w^2}\left[-\beta\cdot\frac{w}{a}\cdot B_f^{\mathrm{II}}\cdot K_f'\left(w\cdot\frac{r}{a}\right)\cos(f\phi+\psi_f)\right.$$

$$\left.+\frac{\omega\varepsilon_2}{r}\,A_f^{\mathrm{II}}\cdot\frac{w}{a}\cdot K_f'\left(w\cdot\frac{r}{a}\right)\cos(f\phi+\alpha_f)\right] \tag{17}$$

$$H_\phi = -\,i\,\frac{a^2}{w^2}\left[\frac{\beta}{r}\cdot B_f^{\mathrm{II}}\cdot K_f\left(w\cdot\frac{r}{a}\right)\cdot f\cdot\sin(f\phi+\psi_f)\right.$$

$$\left. -\,\omega\varepsilon_2\cdot A_f^{\mathrm{II}}\,\frac{w}{a}\cdot K_f'\left(w\cdot\frac{r}{a}\right)\cos(f\phi+\alpha_f)\right] \tag{18}$$

with

$$J_f'(X)=\frac{\mathrm{d}J_f(X)}{\mathrm{d}x}\qquad\text{and}\qquad K_f'(X)=\frac{\mathrm{d}K_f(X)}{\mathrm{d}x} \tag{19}$$

and where α_f and ψ_f are the phase terms associated with the boundary conditions. For $r=a$, continuity of the fields enables one to write

$$A_f^{\mathrm{I}}\cdot J_f(u)=A_f^{\mathrm{II}}\cdot K_f(w) \tag{20}$$

$$\frac{\beta f}{u^2 a}\cdot A_f^{\mathrm{I}}\cdot J_f(u)\sin(f\phi+\alpha_f)+\frac{\omega\mu_0}{ua}\cdot B_f^{\mathrm{I}}\cdot J_f'(u)\cos(f\phi+\psi_f)$$

$$=-\frac{\beta f}{w^2 a}\cdot A_f^{\mathrm{II}}\cdot K_f(w)\sin(f\phi+\alpha_f)-\frac{\omega\mu_0}{wa}\cdot B^{\mathrm{II}}\cdot K_f'(w)\cos(f\phi+\psi_f) \tag{21}$$

$$B_f^{\mathrm{I}}\cdot J_f(u)=B_f^{\mathrm{II}}\cdot K_f(w) \tag{22}$$

$$\frac{\omega\varepsilon_1}{ua}\cdot A_f^{\mathrm{I}}\cdot J_f'(u)\cos(f\phi+\alpha_f)-\frac{\beta f}{u^2 a}\cdot B_f^{\mathrm{I}}\cdot J_f(u)\sin(f\phi+\psi_f)$$

$$=-\frac{\omega\varepsilon^2}{wa}\cdot A_f^{\mathrm{II}}\cdot K_f'(w)\cos(f\phi+\alpha_f)+\frac{\beta f}{w^2 a}\cdot B_f^{\mathrm{II}}\cdot K_f(w)\sin(f\phi+\psi_f) \tag{23}$$

These four equations have solutions for any values of A^{I}, A^{II}, B^{I} and B^{II} on condition that

$$\frac{\omega^2\mu_0}{\beta^2 f^2}\cdot\frac{[(\varepsilon_1/u)(J_f'/J_f)+(\varepsilon_2/w)(K_f'/K_f)]\,[(J_f'/uJ_f)+K_f'/wK_f)]}{[u^{-2}+w^{-2}]^2}$$

$$=-\frac{\sin(f\phi+\psi_f)\sin(f\phi+\alpha_f)}{\cos(f\phi+\psi_f)\cos(f\phi+\alpha_f)} \tag{24}$$

The first term of the equation is independent of the angles ψ_f and α_f. The phases α_f and ψ_f must therefore be considered as independent, and this gives

$$\psi_f - \alpha_f = \pm \frac{\pi}{2} \tag{25}$$

and Equation (24) becomes

$$\left[\frac{J_f'(u)}{uJ_f(u)} + \frac{K_f'(w)}{wK_f(w)} \right] \left[\frac{\varepsilon_1}{\varepsilon_2} \cdot \frac{J_f'(u)}{uJ_f(u)} + \frac{K_f'(w)}{wK_f(w)} \right] = \frac{\beta^2 f^2}{k^2 \varepsilon_2} \left[\frac{1}{u^2} + \frac{1}{w^2} \right]^2 \tag{26}$$

One can now define a normalized frequency parameter V which is proportional to the angular frequency ω of the optical wave and determines the propagation constant of the wave:

$$V^2 = u^2 + w^2 = a^2 k^2 (n_1^2 - n_2^2) = k^2 n_1^2 a^2 2\Delta \tag{27}$$

Hence, starting with a value of the parameter V, it is possible to determine the associated eigenvalues u and w and hence the propagation constant of each mode which can propagate in the fibre in terms of its optoelectronic parameters. In particular, one can write

$$\beta = kn_1 \left[1 - 2\Delta \left(\frac{u}{V} \right)^2 \right]^{1/2} \tag{28}$$

$$\beta = kn_1 \left[1 - \Delta \left(\frac{u}{V} \right)^2 \right] \tag{29}$$

3.2. Classification of modes

Equation (26) is solved by obtaining its eigenvectors, each of which depends on an eigenvalue of the parameter f. A mode which can propagate corresponds to each of these eigenvalues. In particular, the only transverse modes which can propagate without loss are defined by $f = 0$.

3.2.1. TM and TE modes

The modes corresponding to the case $f = 0$ are obtained when one of the terms of (26) is zero. In this case, the first term can be written

$$\frac{\varepsilon_1}{\varepsilon_2} \cdot \frac{J_1(u)}{uJ_0(u)} + \frac{K_1(w)}{wK_0(w)} = 0 \tag{30}$$

which makes the H_r, H_z and E_θ components of the fields equal to zero. These modes are called transverse magnetic ($TM_{0\mu}$) since H_z is zero for $f = 0$; μ is the parameter which defines the μth root of the equation in which case the field equations are:

● for the case $r \leqslant a$:

$$E_z = A_0^{\mathrm{I}} \cdot J_0\left(u \cdot \frac{r}{a}\right) \cos(\alpha_0) \tag{31}$$

$$E_r = \mathrm{i}\, \frac{a}{u} \cdot \beta \cdot A_0^{\mathrm{I}} \cdot J_1\left(u \cdot \frac{r}{a}\right) \cos(\alpha_0) \tag{32}$$

$$H_\phi = \mathrm{i}\, \frac{a}{u} \cdot \omega \varepsilon_1 \cdot A_0^{\mathrm{I}} \cdot J_1\left(u \cdot \frac{r}{a}\right) \cos(\alpha_0) \tag{33}$$

● for the case $r \geqslant a$:

$$E_z = A_0^{\mathrm{I}} \cdot \frac{J_0(u)}{K_0(w)} \cdot K_0\left(w \cdot \frac{r}{a}\right) \cos(\alpha_0) \tag{34}$$

$$E_r = -\mathrm{i}\, \frac{a}{w} \cdot \beta \cdot A_0^{\mathrm{I}} \cdot \frac{J_0(u)}{K_0(w)} \cdot K_1\left(w \cdot \frac{r}{a}\right) \cos(\alpha_0) \tag{35}$$

$$H_\phi = -\mathrm{i}\, \frac{a}{w} \cdot \omega \varepsilon_2 \cdot A_0^{\mathrm{I}} \cdot \frac{J_0(u)}{K_0(w)} \cdot K_1\left(w \cdot \frac{r}{a}\right) \cos(\alpha_0) \tag{36}$$

If the second factor is now made zero, one obtains in the same way

$$\frac{J_1(u)}{u J_0(u)} + \frac{K_1(w)}{w K_0(w)} = 0 \tag{37}$$

which makes the components E_z, E_r and H_ϕ equal to zero. These modes are called transverse electric $(TE_{0\mu})$ with field equations as follows:

● for the case $r \leqslant a$:

$$H_z = B_0^{\mathrm{I}} \cdot J_0\left(u \cdot \frac{r}{a}\right) \cos(\psi_0) \tag{38}$$

$$H_r = \mathrm{i}\, \frac{a}{u} \cdot \beta \cdot B_0^{\mathrm{I}} \cdot J_1\left(u \cdot \frac{r}{a}\right) \cos(\psi_0) \tag{39}$$

$$E_\phi = -\mathrm{i}\, \frac{a}{u} \cdot \omega \mu_0 \cdot B_0^{\mathrm{I}} \cdot J_1\left(u \cdot \frac{r}{a}\right) \cos(\psi_0) \tag{40}$$

● for the case $r > a$:

$$H_z = B_0^{\mathrm{I}} \cdot \frac{J_0(u)}{K_0(w)} \cdot K_0\left(w \cdot \frac{r}{a}\right) \cos(\psi_0) \tag{41}$$

$$H_r = -\mathrm{i}\, \frac{a}{w} \cdot \beta \cdot B_0^{\mathrm{I}} \cdot \frac{J_0(u)}{K_0(w)} \cdot K_1\left(w \cdot \frac{r}{a}\right) \cos(\psi_0) \tag{42}$$

$$E_\phi = \mathrm{i}\, \frac{a}{w} \cdot \omega \mu_0 \cdot B_0^{\mathrm{I}} \cdot \frac{J_0(u)}{K_0(w)} \cdot K_1\left(w \cdot \frac{r}{a}\right) \cos(\psi_0) \tag{43}$$

3.2.2. Hybrid modes

In the case where $f \geqslant 1$, one can no longer have clear-cut cases since the components E_z and H_z are no longer zero; hence the name hybrid mode. In these fibres, the refractive indices of the core and cladding are only slightly different hence $\varepsilon_1 \approx \varepsilon_2$ and the characteristic equation (37) can be written

$$\frac{J_f'(u)}{uJ_f(u)} + \frac{K_f'(w)}{wK_f(w)} = \pm f[u^{-2} + w^{-2}] \tag{44}$$

The + sign defines fields such that $H_z > E_z$ and the − sign such that $E_z > H_z$. By applying the same reasoning used previously, the boundary condition for $r = a$ defines the parameter P for hybrid modes:

$$P = -\frac{\mu_0\omega}{\beta} \cdot \frac{B_f^1}{A_f^1} \cdot \frac{\cos(f\phi + \psi_f)}{\sin(f\phi + \alpha_f)}$$

$$= \frac{f[u^{-2} + w^{-2}]}{[[J_f'(u)/uJ_f(u)] + [K_f'(w)/wK_f(w)]]} \tag{45}$$

By putting

$$Fc = A_f^1 \cdot \exp[i(\omega t - \beta z)] \cdot \cos(f\phi + \alpha_f)$$
$$Fs = A_f^1 \cdot \exp[i(\omega t - \beta z)] \cdot \sin(f\phi + \alpha_f) \tag{46}$$

one obtains the field equations:

● for the case $r \leqslant a$

$$E_z = J_f\left(u \cdot \frac{r}{a}\right)Fc \tag{47}$$

$$H_z = -\frac{\beta \cdot P}{\mu_0\omega} \cdot J_f\left(u \cdot \frac{r}{a}\right)Fs \tag{48}$$

$$E_r = -i\beta \cdot \frac{a}{u}\left[J_f'\left(u \cdot \frac{r}{a}\right) - P \cdot \frac{f}{r} \cdot \frac{a}{u} \cdot J_f\left(u \cdot \frac{r}{a}\right)\right]Fc \tag{49}$$

$$E_\phi = i\beta \cdot \frac{a}{u}\left[\frac{f}{r} \cdot \frac{a}{u} \cdot J_f\left(u \cdot \frac{r}{a}\right) - P \cdot J_f'\left(u \cdot \frac{r}{a}\right)\right]Fs \tag{50}$$

$$H_r = i\frac{k^2n_1^2a}{\omega\mu_0 u}\left[P \cdot \frac{\beta^2}{k^2n_1^2} \cdot J_f'\left(u \cdot \frac{r}{a}\right) - \frac{f}{u} \cdot \frac{a}{r} \cdot J_f\left(u \cdot \frac{r}{a}\right)\right]Fs \tag{51}$$

$$H_\phi = i\frac{k^2n_1^2a}{\omega\mu_0 u}\left[P \cdot \frac{\beta^2}{k^2n_1^2} \cdot \frac{f}{u} \cdot \frac{a}{r} \cdot J_f\left(u \cdot \frac{r}{a}\right) - J_f'\left(u \cdot \frac{r}{a}\right)\right]Fc \tag{52}$$

with the additional relation

$$B_f^I = \frac{\beta}{k} \left[\frac{\varepsilon_0}{\mu_0}\right]^{1/2} \cdot P \cdot A_f^I \tag{53}$$

● for the case $r > a$

$$E_z = J_f(u) \cdot \frac{K_f(w \cdot r/a)}{K_f(w)} \cdot Fc \tag{54}$$

$$H_z = -\frac{\beta \cdot P}{\mu_0 \omega} \cdot J_f(u) \cdot \frac{K_f(w \cdot r/a)}{K_f(w)} \cdot Fs \tag{55}$$

$$E_r = -i\beta \cdot \frac{a}{w} \left[-J_f(u) \cdot \frac{K_f(w \cdot r/a)}{K_f(w)} + P \cdot \frac{f}{r} \cdot \frac{a}{w} \cdot J_f(u) \cdot \frac{K_f(w \cdot r/a)}{K_f(w)} \right] Fc \tag{56}$$

$$E_\phi = -i\beta \cdot \frac{a}{w} \left[\frac{f}{r} \cdot \frac{a}{w} J_f(u) \cdot \frac{K_f(w \cdot r/a)}{K_f(w)} - P \cdot J_f(u) \cdot \frac{K_f(w \cdot r/a)}{K_f(w)} \right] Fs \tag{57}$$

$$H_r = i \frac{k^2 n_2^2 a}{\omega \mu_0 w} \left[-P \cdot \frac{\beta^2}{k^2 n_2^2} J_f(u) \cdot \frac{K_f'(w \cdot r/a)}{K_f(w)} \right.$$
$$\left. + \frac{f}{w} \cdot \frac{a}{r} \cdot J_f(u) \cdot \frac{K_f(w \cdot r/a)}{K_f(w)} \right] Fs \tag{58}$$

$$H_\phi = i \frac{k^2 n_2^2 a}{\omega \mu_0 w} \left[-P \cdot \frac{\beta^2}{k^2 n_2^2} \cdot \frac{f}{w} \cdot a J_f(u) \cdot \frac{K_f(w \cdot r/a)}{K_f(w)} + J_f(u) \cdot \frac{K_f(w \cdot r/a)}{K_f(w)} \right] Fc \tag{59}$$

with the additional relations:

$$A_f^{II} = A_f^I \cdot \frac{J_f(u)}{K_f(w)} \tag{60}$$

$$B_f^{II} = B_f^I \cdot \frac{J_f(u)}{K_f(w)} = \frac{\beta}{k} \cdot \left[\frac{\varepsilon_0}{\mu_0}\right]^{1/2} \cdot P \cdot \frac{J_f(u)}{K_f(w)} \cdot A_f^I \tag{61}$$

In this way, expansion of Maxwell's equations enables expressions for the field components of each mode which propagates in the fibre to be obtained and the various types of mode which exist to be determined. The computations have been developed for a step index fibre but can also be developed for a graded index fibre.

4. CUTOFF FREQUENCY

The expressions for the fields in the optical cladding are described by a function $K_c(wr/a)$ which decreases very rapidly with r as long as $\beta > k n_2$ and means that the mode is confined to the core. When β tends to kn_2 the mode field extends more and more into the cladding until it is no longer guided. This limiting value corresponds to the cut-off frequency and one has

$$w = 0 \qquad V = V_c = u_c \tag{62}$$

A limited expansion of the expression for $K_f(x)$ gives $(x \approx 0)$:

$$
\begin{aligned}
K_0(x) &\approx -\log(x) & f = 0 \\
K_f(x) &\approx \frac{2^{f-1}}{x^f} \cdot (f-1)! & f \geqslant 1
\end{aligned}
\tag{63}
$$

4.1. The case of transverse modes

Retaining the previous assumptions, one has

$$\frac{J_1(u)}{u J_0(u)} = \frac{1}{w^2 \log(w)} \tag{64}$$

In the case $w = 0$, eigenvalue equations (37) and (44) reduce to

$$J_0(u_c) = 0 \tag{65}$$

and the fields in the cladding $(r > a)$ for the transverse magnetic case $(TM_{0\mu})$ can be written:

$$E_z = 0 \tag{66}$$

$$E_r = i \frac{a^2}{u_c} \cdot \frac{\varepsilon_1}{\varepsilon_2} \cdot J_1(u_c) \cdot kn_2 \frac{A_0^I}{r} \cos(\alpha_0) \tag{67}$$

$$H_\phi = i\omega \cdot a^2 \cdot \frac{\varepsilon_1}{u_c} J_1(u_c) \frac{A_0^I}{r} \cos(\alpha_0) \tag{68}$$

for the transverse electric fields $(TE_{0\mu})$:

$$H_z = 0 \tag{69}$$

$$H_r = i \frac{a^2}{u_c} \cdot \beta \cdot J_1(u_c) \cdot kn_2 \frac{B_0^I}{r} \cos(\psi_0) \tag{70}$$

$$E_\phi = -i\omega \cdot \mu_0 \cdot \frac{a^2}{u_c} J_1(u_c) \frac{B_0^I}{r} \cos(\psi_0) \tag{71}$$

which correspond to an electromagnetic wave decreasing as $1/r$ in the cladding.

4.2. The case of hybrid modes

In the case of hybrid modes, the characteristic equation (44) has a cut-off frequency given by

$$2f \cdot u_c \cdot J_f(u_c) \left[J_{f-1}(u_c) \left[\frac{\varepsilon_1}{\varepsilon_2} + 1 \right] - \frac{u_c J_f(u_c)}{(f-1)} \right] = 0 \qquad (72)$$

For the $f = 1$ mode, this equation has a solution only for $u_c = 0$ which is the case of the HE_{11} modes for which the low-frequency cut-off is zero. the $HE_{f\mu}$ modes have a cut-off frequency which is the μth root of the equation:

$$(f-1)\left(\frac{\varepsilon_1}{\varepsilon_2} + 1\right) J_{f-1}(u_c) = u_c \cdot J_f(u_c)$$

which reduces to:

$$J_{f-2}(u_c) = 0 \qquad \text{if } \varepsilon_1 \approx \varepsilon_2$$

The $EH_{f\mu}$ modes have a cutoff frequency which corresponds to the μth root of:

$$J_f(u_c) = 0$$

For the $f > 1$ modes, a solution cannot be obtained with $u_c = 0$ and $w = 0$ simultaneously but there is always a compromise between the cladding radius (a) and the index Δ to ensure a region of guided propagation without modal loss.

5. DEGENERACY OF MODES

Again using the equation for the eigenvalues of the $HE_{f\mu}$ and $EH_{f\mu}$ modes, one has

$$\left[\frac{J_{f-1}(u)}{f J_f(u)} - \frac{K_{f-1}(w)}{w K_f(w)} \right] \left[\frac{J_{f+1}(u)}{u J_f(u)} + \frac{K_{f+1}(w)}{w K_f(w)} \right] = 0 \qquad (73)$$

Each of the terms can be equal to zero in order to give the eigenvalues which define the modes which propagate:

$$u \cdot \frac{J_f(u)}{J_{f+1}(u)} = \pm w \cdot \frac{K_f(u)}{K_{f+1}(w)} \qquad (74)$$

By using the properties of Bessel functions, this equation can be written in the form:

$$u \cdot \frac{J_{m-1}(u)}{J_m(u)} = -w \cdot \frac{K_{m-1}(w)}{K_m(w)} \qquad (75)$$

in which for

$$m = 1, \quad \text{the modes are } TE \text{ or } TM$$
$$m = f + 1, \quad \text{the modes are } EH_{f\mu}$$
$$m = f - 1, \quad \text{the modes are } HE_{f\mu}$$

Thus all the modes defined by a common pair of values of m and μ satisfy the same equation for the eigenvalues. This leads to a degeneracy of modes and indicates that all modes have a common phase constant which leads to the concept of linearly polarized modes.

6. MONOMODE PROPAGATION CONDITIONS

It has been observed that only the HE_{11} mode has a low cut-off frequency of zero. The modes obtained for $m = 1$, such as TE_{01}, TM_{01} and HE_{21} have a non-zero cutoff frequency $V_c = 2.4048$. It is thus possible to find opto-electronic parameters such that

$$V < 2.4048 = \frac{2\pi a}{\lambda} \cdot n_1 \cdot (2\Delta)^{1/2} \tag{76}$$

which permit propagation of a monomode wave, and a wavelength exists beyond which propagation will be monomode:

$$\lambda_c = 3.695 \cdot a \cdot n_1 \cdot (\Delta)^{1/2} \tag{77}$$

The most used are 1.3 μm and 1.5 μm. In the same way, for wave propagation to occur in the core, the core must satisfy the relation

$$a < 0.38 \cdot \frac{\lambda_c}{(n_1^2 - n_2^2)^{1/2}} \tag{78}$$

7. THE CONFINEMENT FACTOR

7.1. Definition

The energy transported by an electromagnetic wave is given by the Poynting vector S, for which

$$S = E \wedge H \tag{79}$$

The total flux flowing through a surface A with a normal n is

$$P \int_A S \cdot n \, dA \tag{80}$$

It is possible to decompose the transverse components of a field on a reference plane xOy such that

$$E_x = E_r \cdot \cos(\phi) - E_\phi \cdot \sin(\phi) \tag{81}$$

$$E_y = E_r \cdot \sin(\phi) + E_\phi \cdot \cos(\phi) \tag{82}$$

$$H_x = H_r \cdot \cos(\phi) - H_\phi \cdot \sin(\phi) \tag{83}$$

$$H_y = H_r \cdot \sin(\phi) + H_\phi \cdot \cos(\phi) \tag{84}$$

in this case one has

$$S_z = E_x \cdot H_y - E_y \cdot H_x \tag{85}$$

for which the expression averaged over time is written:

$$\langle S_z \rangle = \frac{n_1 \cdot k}{2\omega\mu_0} \cdot E_x^2 = \frac{n_1}{2Z_0} \cdot E_x^2 \tag{86}$$

where Z_0 is the impedance of free space. The optical power confined in the core can be written as

$$P_c = \frac{1}{2} \cdot \frac{n_1}{Z_0} \cdot E_m^2 \int_0^a \int_0^{2\pi} \left[\frac{J_m(u \cdot r/a)}{J_m(u)} \right]^2 \cos^2(m\phi) \cdot r \, dr \, d\phi \tag{87}$$

and that confined in the cladding

$$P_{cl} = \frac{1}{2} \cdot \frac{n_2}{Z_0} \cdot E_m^2 \int_0^a \int_0^{2\pi} \left[\frac{K_m(w \cdot r/a)}{K_m(w)} \right]^2 \cos^2(m\phi) r \, dr \, d\phi \tag{88}$$

where

$$E_m = \beta \cdot \frac{a}{u} \cdot J_m(u) \tag{89}$$

which defines the amplitude of the field at the interface $r = a$. The total power propagating in the fibre for the $m = 0$ modes can be written:

$$P_T = \frac{\pi a^2 E_0^2}{2Z_0} \left[n_1 \left(1 + \left(\frac{J_1(u)}{J_0(u)} \right)^2 \right) + n_2 \left(\left(\frac{K_1(w)}{K_0(w)} \right)^2 - 1 \right) \right] \tag{90}$$

The confinement factor is defined as the ratio of the power in the core and the total power which for the $m = 0$ modes can be written:

$$F_0 = \frac{P_c}{P_T} = \frac{w^2}{V^2} \left[1 + \left(\frac{J_0(u)}{J_1(u)} \right)^2 \right] \tag{91}$$

It should be noted that at the low cut-off of the fundamental mode $V = 2.4048$, 82.5% of the energy is confined in the core of a step index fibre.

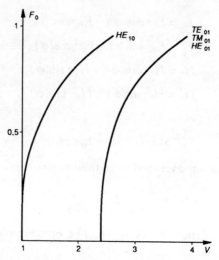

Figure 3.3 Confinement factor as a function of the parameter V

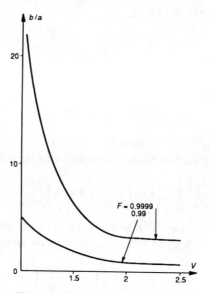

Figure 3.4 Radial value as a function of V

7.2. Radial extent of the energy

The energy contained within a radius b of the fibre can be written

$$P_{ext} = \int_1^R [K_0(wR)]^2 R \, dR = \frac{1}{2} \left[\left(F_0\left(\frac{V^2}{w^2}\right) - 1 \right) \frac{w^2}{u^2} K_1^2(w) - K_0^2(w) \right] \quad (92)$$

where $R = r/a$; it defines the radial extent of the energy in the fibre.

Monomode optical fibres used in telecommunications have modes which occupy a radius $b \leqslant 4a$; this value defines the optical cladding before deposition in order to produce monomode optical fibres.

7.3. The normalized propagation parameter

Let this parameter be B which is defined by

$$B = 1 - \frac{u^2}{V^2} = \frac{(\beta/k)^2 - n_2^2}{n_1^2 - n_2^2} \quad (93)$$

Substituting this expression in the propagation constant, one obtains

$$\beta \approx kn_2(1 + B\Delta) = kn_2\left[1 + \Delta\left(1 - \frac{u^2}{V^2}\right)\right] \quad (94)$$

This relation permits simple calculation of the propagation constant of a

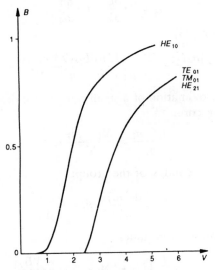

Figure 3.5 Variation of the normalized propagation constant as a function of the group order

mode as a function of Δ and V. This parameter arises in the determination of the amplitude of dispersion of the mode in the guiding structure.

8. SPECTRAL ATTENUATION

Several physical phenomena contribute to spectral attenuation, which arises from the various losses in the fibre. During fabrication of fibres an attempt is made to minimize them in order to increase the distance between repeaters.

Firstly, losses by absorption in the ultra-violet and infra-red can be identified. The first are associated with the existence of local macroscopic electric fields in the glass; the second arise from the existence of different vibrational modes of the tetrahedral structure of silica (SiO_2). Diffusion losses are also found due to fluctuations of density or doping in the fibre; these lead to local variations of refractive index. It is also necessary to take account of the influence of hydroxide ions which cause an absorption band in the region of 0.9 and 1.5 μm.

Detailed analysis shows that the theoretical lower limit, due to Rayleigh scattering, which can be obtained is 0.16 dB/km at 1.54 μm for the best case.

9. DISPERSION IN MONOMODE FIBRES

The mode group propagation delay in the fibre is determined by

$$\tau_g = \frac{L}{c} \cdot \frac{d\beta}{dk} = L \cdot \frac{d\beta}{d\omega} \qquad (95)$$

hence using Equation (93)

$$\tau_g = \frac{L}{c} \cdot \frac{[n_2 N_2 + (B + V/2)(dB/dV)(n_1 N_1 - n_2 N_2)]}{[n_2^2 + (n_1^2 - n_2^2)B]^{1/2}} \qquad (96)$$

By applying the approximation of a small refractive index difference between the cladding and the core,

$$\frac{n_1 - n_2}{n_2} \approx \frac{N_1 - N_2}{N_2} \ll 1 \qquad (97)$$

where N_i represents the index of the group:

$$N_i = \frac{d(kn_i)}{dk} = n_i - \lambda \frac{dn_i}{d\lambda} \qquad (98)$$

the group delay expression becomes

$$\tau_g = \frac{L}{c} \left[N_2 + (N_1 - N_2) \frac{d(VB)}{dV} \right] \qquad (99)$$

The first term (N_2) determines the dispersion of the material and the second the dispersion of the mode due to guiding.

If one wishes to examine the influence of a spectral component on the group delay and account is taken of a particular width of this component, one has, to a second-order approximation

$$\frac{1}{L} \cdot \frac{d\tau_g}{d\lambda} = M_1 D + (1 - D) \cdot M_2 - \frac{n_2 \Delta}{\lambda c} \cdot V \cdot \frac{d^2(VB)}{dV^2}$$

$$- \frac{n_2}{\lambda c} \cdot \frac{d(\Delta)}{d\lambda} \cdot \left[V \cdot \frac{d^2(VB)}{dV^2} + \frac{d(VB)}{dV} - B \right] \quad (100)$$

putting

$$D = \frac{1}{2} \left(B + \frac{d(VB)}{dV} \right) \quad (101)$$

$$\Delta \approx \frac{n_1 - n_2}{n_2} \quad (102)$$

$$M_i = \frac{1}{c} \cdot \frac{dN_i}{d\lambda} = -\frac{\lambda}{c} \cdot \frac{d^2 n_i}{d\lambda^2} \quad (103)$$

The first two terms represent the dispersion due to the material, the third defines the dispersion due to modes in the guide and the fourth defines the variation of refractive index with wavelength.

For core and cladding refractive indices which are close, one has

$$\frac{1}{L} \cdot \frac{d\tau_g}{d\lambda} = M_2 + (M_1 - M_2) \cdot \frac{d(VB)}{dV} - \frac{1}{c\lambda} \cdot (N_1 - N_2) \cdot \frac{N_2}{n_2} V \cdot \frac{d^2(VB)}{dV^2} \quad (104)$$

It should be noted that the dispersion is virtually zero when $M_1 \approx M_2$ on condition that the refractive index difference satisfies

$$\Delta n \geqslant \frac{M_2 \lambda_c}{V [d^2(VB)/dV^2]} \quad (105)$$

Making use of the fact that the two factors of (104) are of opposite signs, it can be arranged that one compensates for the other on condition that they have approximately the same amplitude. This amounts to compensating for the material dispersion by means of modal dispersion.

10. MULTIDIELECTRIC STRUCTURES

The use of vapour phase deposition techniques permits the production of monomode fibres which have more complex refractive index profiles; the most common is the W type.

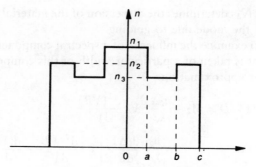

Figure 3.6 W-type refractive index profile

The advantage of these fibres is that they have a modal dispersion which is sufficiently large to compensate for the material dispersion over a wide spectral range; this reduces the overall losses and enables the bandwidth and hence transmission capacity of the fibre to be increased. The same type of mathematical analysis is applicable to these structures in order to determine the modes which propagate.

11. THE PERFORMANCE OF MONOMODE OPTICAL FIBRES

Monomode optical fibres usually have a W type refractive index profile (see Figure 3.6), the performance depends on the wavelength used and the principal characteristics are given in Table 3.1.

These high-performance monomode fibres have serious operational difficulties which are associated with the small core diameter ($\simeq$ 6 or 7 μm). The cores must be aligned at the various interfaces (such as transmitters, connectors and receivers) to better than 1 μm in order to minimize coupling losses.

Table 3.1 The performance of some monomode optical fibres

Core diameter	7 μm
Optical cladding diameter	42 μm
Δn	6.5×10^{-3}
Attenuation at 1.3 μm	0.43 dB/km
Attenuation at 1.55 μm	0.29 dB/km
Capacity	> GHz/km

4

FABRICATION OF OPTICAL FIBRES

1. THE PRINCIPLES OF FABRICATION

The performance of optical fibres depends very much on the fabrication processes and the materials used; there are two objectives as follows:

— the production of very pure glass;
— precise control of the refractive index profile in graded index optical fibres.

The high grade silica glasses used for the production of optical fibres have a viscosity which makes the use of traditional glass-making methods to obtain a very low level of impurities difficult. Particles of SiO_2 glass are therefore produced by chemical reactions for which the basis is

$$SiCl_4 + O_2 \leftrightarrow SiO_2 + 2Cl_2$$

It is possible to vary the refractive index of the material by mixing dopants with the basic products of the chemical reaction. The difficulty lies in the control of the refractive index during fabrication to obtain an optimum value; the chemical reactions are as follows:

$$GeCl_4 + O_2 \leftrightarrow GeO_2 + 2Cl_2$$
$$2POCl_3 + 3/2O_2 \leftrightarrow P_2O_5 + 3Cl_2$$
$$2BBr_3 + 3/2O_2 \leftrightarrow B_2O_3 + 3Br_2$$

The first step in obtaining an optical fibre is fabrication of a bar of glass called a preform which consists of a core and an optical cladding. The diameter of preforms can vary from 1 to 10 or more centimetres according to the technique used; their length is usually 1 metre. The preform is then drawn to produce a fibre; this is fibre drawing.

Fabrication of the preform permits certain parameters to be determined as follows:

— the optoelectronic characteristics of the fibre;
— the uniformity of the core–cladding ratio;
— the circularity of the optical fibre.

In contrast, the uniformity of the external diameter and the mechanical resistance of the fibre depend on the fibre-drawing operation.

2. FABRICATION OF PREFORMS

2.1. Fibre materials

Fibres consist of an optical cladding of $B_2O_2-P_2O_5-SiO_2$ and a core of $GeO_2-P_2O_5-SiO_2$. The dopants which can be used to increase the refractive index are germanium and phosphorus; those which enable the index to be reduced are boron and fluorine.

2.2. The MCVD method

The preforms are fabricated in two successive stages which must be performed without removal and without interruption to avoid thermal shocks which cause discontinuities in the refractive index.

2.2.1. Deposition

A tube of silica mounted in a glass lathe is heated to a high temperature (1400–1600°C) by means of an oxyhydrogen blowtorch which moves parallel to the tube over the whole of the available length. At one end of the tube, a gaseous mixture of oxygen and halides is injected. This mixture reacts in the heated region to produce a mixture of finely divided oxides which is deposited in powder form slightly downstream on the internal wall. This deposit is then vitrified by the passage of the blowtorch.

The chemical relations depend on the temperature and require it to be precisely controlled; once it is set, there is a thickness limit of the initial layer of the order of 10–20 μm. It is thus necessary to deposit many layers by making a large number of passes of the blowtorch. In this way multilayer graded index fibres are obtained.

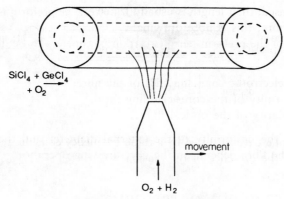

Figure 4.1   Deposition by MCVD

2.2.2. Collapsing

When the deposit is complete, injection of gas is stopped and the temperature is increased to 2000°C. At this temperature, the silica softens sufficiently for the forces, due to the high surface temperature, to produce a homogeneous necking of the tube. The necking becomes total after 3–5 passes of the blowtorch; this produces a glass rod or preform ready for fibre drawing.

2.3. The PMCVD method

The plasma modified chemical vapour deposit (PMCVD) method is similar to the MCVD process but a low-pressure isothermal plasma is introduced into

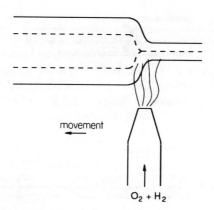

Figure 4.2   The principle of collapsing

Fabrication of optical fibres

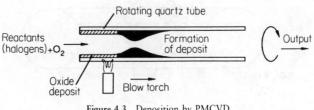

Figure 4.3   Deposition by PMCVD

the tube within a cavity excited by a radio-frequency coil; the reaction is thus stimulated by the presence of the plasma.

2.4. The PCVD method

The plasma chemical vapour deposit (PCVD) method is similar to the PMCVD process except that the radio-frequency coil for exciting the plasma is replaced by a microwave oven.

2.5. The OVPD method

The outside vapour phase deposit (OVPD) process consists of depositing the doped silica on an aluminium oxide mandrel; the graded index is obtained by modifying the constitution of the gaseous mixture.

After vitrifying the deposit, the central mandrel is removed and the tube is collapsed to obtain the preform.

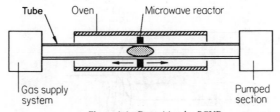

Figure 4.4   Deposition by PCVD

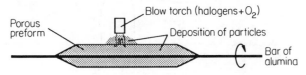

Figure 4.5   Deposition by OVPD

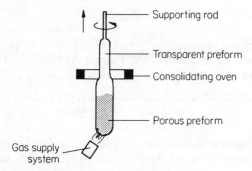

Figure 4.6 Fabrication of a preform by the VAD process

2.6. The VAD process

The vapour axial deposit (VAD) process permits axial increase of the preform; the index profile is obtained by radial control of the concentrations.

The bar is then consolidated in a controlled atmosphere to produce a homogeneous vitreous preform.

2.7. The ALPD process

The axial lateral plasma deposition (ALPD) process has been developed by Saint Gobain and is performed in two stages as follows:

— Fabrication in a plasma torch of a core preform which can be of graded or step index.
— Deposition of the cladding around the core preform.

Consequently no collapsing operation is required.

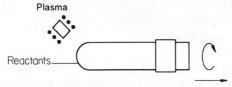

Figure 4.7 Fabrication of a core preform by ALPD

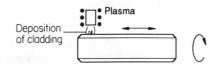

Figure 4.8 Fabrication of the cladding on the core preform

2.8. Comparison of the methods

The MCVD method permits realization of all profiles; the VAD technique enables the central dip to be avoided and the ALPD, OVPD and VAD techniques allow preforms to be obtained without using an initial tube of silica.

Process	MCVD	VAD	ALPD
Speed of deposition (g/mm)	0.4–2.3	0.4–4.5	3–10
Quantity of fibre per preform (km)	8–40	13–40	100–250

These processes allow the following to be obtained in industrial production:

— generally good reproducibility
— very low attentuation:
  < 3 dB/km at 820 nm
  < 1 dB/km at 1300 nm;
— high bandwidths:
  up to 800 MHz at 820 nm
  up to 15 GHz at 1300 nm;
— high mechanical resistance:
  up to 4 GPa.

2.9. The hardware

Systems for producing preforms contain the following three essential components:

— a glass lathe;
— a gas distribution system;
— a control system.

2.9.1. The glass lathe

This is a conventional horizontal lathe which includes a slide which moves horizontally and is equipped with an optical pyrometer temperature transducer. The crucial factor is the stability of the speed of movement which ensures good homogeneity of the deposit.

2.9.2. The gas distribution system

This consists of electric valves and flowmeters which provide and control all the gas used. The critical factor is regulation of the gas flow.

2.9.3. The control system

This is generally a computer system which controls and monitors the following important parameters:

— the speed of movement of the blowtorch;
— the external temperature of the silica tube;
— the oxygen flow rate;
— the flow rates of the various halides;
— opening and closing of the valves;
— alarms.

2.10 Fibre drawing

Optical fibre is obtained by drawing from the preform at high temperature. This operation, which must conserve the qualities of the preform, is immediately followed by an induction operation which surrounds the optical fibre with a protective coating.

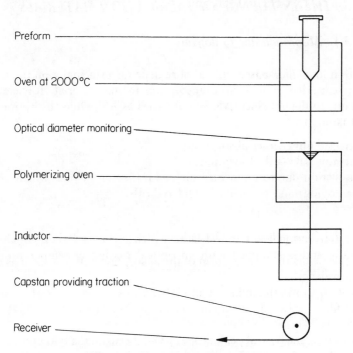

Preform

Oven at 2000°C

Optical diameter monitoring

Polymerizing oven

Inductor

Capstan providing traction

Receiver

Figure 4.9   The principle of fibre drawing

Drawing of an optical fibre is controlled solely by the speed of the drawing capstan. Continuous measurement of the diameter of the optical fibre enables slow variations of thickness due to inevitable temperature fluctuations to be avoided by controlling the speed of the capstan in accordance with the diameter of the optical fibre. Consequently, the precision of the diameter of the optical fibre is typically $\pm 0.1\ \mu$m.

The products used in the inductor are essentially silicones or epoxyacrylate resins which polymerize as they pass through the oven; the purpose of this layer is to ensure protection of the optical fibre and to increase its mechanical strength.

The main parameters which are controlled in a fibre drawing system are as follows:

— the stability of the supports;
— the constancy of the drawing speed;
— the stability of the temperatures;
— the constancy of the diameter of the optical fibre.

3. THE INFLUENCE OF FABRICATION MATERIALS

3.1. Absorption due to dopants

Attenuation in a silica-based optical fibre depends principally on the intrinsic Rayleigh scattering and intrinsic absorption in the material. Rayleigh scattering is due to density fluctuations and varies as λ^{-4} while absorption in the material is due to:

— intrinsic ultra-violet absorption;
— intrinsic infra-red absorption;
— absorption due to impurities such as transition metals (which are used for doping) and particularly OH radicals.

3.1.1. The influence of OH radicals

OH radicals define the general form of the absorption curve of silica as a function of wavelength.

Figure 4.10 shows the influence of OH radicals on the spectral attenuation of a silica fibre for two different concentrations (the first is 0.1×10^{-6} and the second 5×10^{-6}).

In both cases, the central peak is at 1.39 μm between the two minima which are normally used for optical telecommunication (1.3 μm and 1.5 μm); its

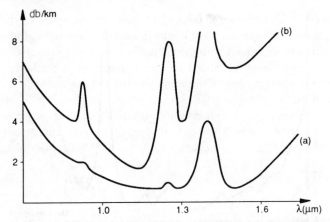

Figure 4.10   Spectral attenuation for two concentrations of OH (curve (a) < 0.1 and curve (b) ≈ 5×10^{-6})

maximum reaches 38 dB for the higher concentration. All the peaks visible on this curve are associated with the absorption bands of OH radicals and not silica; this shows their importance and the attention which must be devoted to them in fibre fabrication.

3.1.2. Sources of OH radicals

Three main sources of OH radicals can be identified:

(a) The oxygen used, although purified, always has a residue of OH radicals.
(b) The collapsing operation, which is very difficult, is extremely sensitive to contamination by OH radicals due to backward diffusion of air at the time of the operation.
(c) The joints used in the gas piping have some porosity to these radicals.

3.2. The influence of structural defects

The presence of structural defects is caused by inadequate cleaning of the silica tube (CVD method), the return of deposits in the absence of heating and the nature of the dopants. The main cause is the nature of the interface between the silica preform mould and the deposit. Bubbles form at this interface where there are surface impairments; these increase the Rayleigh scattering and hence the attentuation.

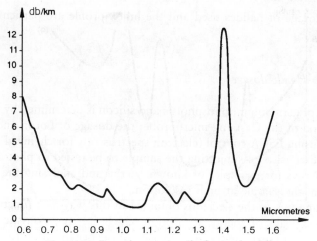

Figure 4.11 Spectral attenuation of a fluorine doped fibre

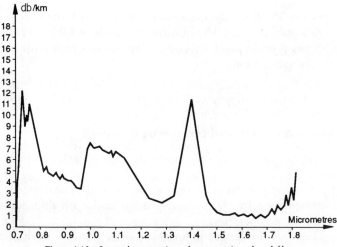

Figure 4.12 Spectral attenuation of a germanium doped fibre

3.3. The influence of dopants

The spectral attenuation curve varies in accordance with the dopants used; these are principally fluorine and germanium.

4. CONTROL OF THE FABRICATION PROCESS

Research has shown that the optoelectronic parameters of the preform are reproduced, with scaling, in the fibre. During fabrication of the preform, the dosage of halides used and the index profile are particularly well controlled.

4.1. Halide dosage

The dosage of germanium, phosphorus and silicon is determined by the traditional method of the Castaing microprobe; the dosage of boron and fluorine is achieved using ESCA control (electron spectroscopy for chemical analysis). This method consists of subjecting the sample to be tested to photon irradiation (soft X-rays for example) of known energy and analysing the electrons emitted from the sample at high resolution.

Spectral analysis of the electrons obtained in this way as a function of the energy shows peaks of energy absorption which are characteristic of a particular substance. Furthermore, the intensity of the peak is proportional to the concentration in atoms of the element which provides the electrons, hence the quantitative aspect.

4.2. The refractive index profile

In general, the actual index profile always has a central hollow due to the evaporation of dopants during the collapsing operation. This hollow influences the propagation of some modes.

Of the measurement methods used, two are based on the deviation of light rays by a transversely illuminated preform; they have the advantage of being non-destructive and applicable to any fabrication process. The exception is the focusing method which requires a central hollow in the index to facilitate adjustment.

In all cases, the preform is immersed in an index matching liquid and is compared with a cylindrical lens whose refractive index varies in the core region.

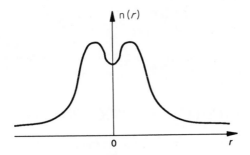

Figure 4.13   An actual refractive index profile

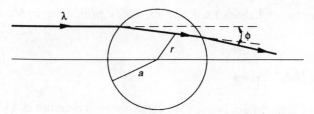

Figure 4.14   Deviation of a beam by the core

The direct method consists of sending a ray through the preform and analysing its deviation at the output (ϕ); this is directly proportional to the height of incidence and the refractive index.

The focusing method consists of measuring the energy distribution in the plane of observation which itself depends on the height of incidence of the beam and the refractive index.

5

CHARACTERIZATION OF OPTICAL FIBRES

1. INTRODUCTION

Several parameters enable the qualities of an optical fibre to be defined:

— the refractive index profile;
— the numerical aperture;
— the diameter;
— the bandwidth and associated pulse spreading and frequency dispersion;
— losses;
— modal dispersion;
— spectral attenuation.

For each of these parameters, several methods have been developed which enable them to be measured with varying degrees of accuracy. They are indispensable to the analysis of fibre optic links.

2. NUMERICAL APERTURE

This is one of the parameters to be measured since it is required for all measurement methods which use injection of light into the optical fibre.

It has been seen that the numerical aperture (NA) of an optical fibre is defined by:

$$NA = \frac{(n_1^2 - n_2^2)^{1/2}}{n_0} \qquad (1)$$

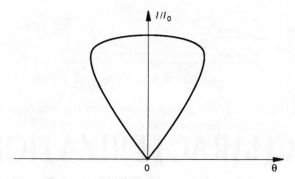

Figure 5.1   Radiation diagram of an optical fibre

In the case where index matching fluid is not used for injection of the light, $n_0 = 1$. In addition to this theoretical maximum aperture, an effective numerical aperture is defined as follows:

$$NA_{eff} = \sin(\theta_m)$$

where m is the half angle at the vertex of the cone containing 90% of the energy. This parameter depends on the excitation conditions and the length of the optical fibre.

Measurement is made by plotting the far field ray diagram and then determining the angle of the cone containing 90% of the energy. To obtain a balance of the modes in the optical fibre, a mode mixer is used which is inserted between the optical fibre to be tested and the light injecting optics.

3. MEASUREMENT OF REFRACTIVE INDEX PROFILE

3.1. The near field method

3.1.1. The principle

This simple method enables the refractive index profile to be measured by measuring the luminous power density emerging from a multimode or monomode optical fibre when it is illuminated with a Lambertian source. The principle consists of sending light through an optical fibre and observing the distribution of the light emerging from the fibre with a microscope coupled to a detector.

On entering the optical fibre, the light is subjected to refraction such that

$$n_0 \cdot \sin(\theta_0) = n(r) \cdot \sin(\theta_f) \tag{2}$$

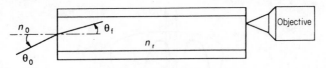

Figure 5.2 The basis of observation

In the optical fibre, only the guided modes reach the other end and this corresponds to a maximum propagation angle θ_c. Ignoring both leaky modes and absorption losses, the light distribution can be considered to be independent of z if the source is Lambertian, in which case

$$n(r)\cos(\theta_f) = \frac{b}{k} \tag{3}$$

for any light ray; $\beta = kn_2$ corresponds to the maximum propagation constant of a wave which can propagate:

$$n(r)\cos(\theta_c) = n_2 \tag{4}$$

and this relates the maximum angle to the refractive index of the cladding.

Now consider a Lambertian source; the power emitted in a solid angle $d\Omega$ by a surface element dS is:

$$dP = I_0 \cdot \cos(\theta) \cdot d\Omega \cdot dS$$

As $d\Omega = \sin(\theta) \cdot d\theta \cdot d\psi$

$$dP = I_0 \cdot \cos(\theta) \cdot \sin(\theta) \cdot d\theta \cdot d\psi \cdot dS$$

$$dP = I_0 \cdot \frac{n^2(r)}{2n_0^2} \cdot d(\sin^2(\theta_f)) \cdot d\theta \cdot d\psi \cdot dS \tag{5}$$

$$P_d(r) = \frac{I_0 \cdot S_d \cdot n^2(r)}{2n_0^2} \cdot \int_0^{2\pi} d\psi \cdot \int_0^{\sin^2\theta_c} d(\sin^2(\theta_f)) \tag{6}$$

$$P_d(r) = \frac{I_0 \cdot S_d}{2n_0^2} \cdot \pi \cdot [n^2(r) - n_2] \tag{7}$$

By displacing the point of convergence along a diameter of the fibre, a plot of the refractive index profile is obtained.

In order to increase the accuracy of measurement, the light is modulated and synchronous detection is used. The central hollow associated with collapsing the glass during fabrication is again evident.

3.1.2. Equipment

The equipment (Figure 5.4) consists of a short optical fibre to avoid losses due to mode coupling; the Lambertian source is injected through a focusing lens.

Characterization of optical fibres

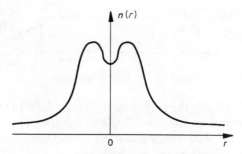

Figure 5.3 Measured refractive index profile

The receiving objective is combined with a lens in order to obtain a parallel beam. This beam is chopped and received by a photodiode which moves in the image plane. The reference signal from the modulator and that from the detector are fed to a comparator and then to a plotter.

3.1.3. Limitations of the method

To avoid mode coupling losses and differential losses, very short optical fibres are used (less than 1 metre). With this method it is not possible to measure the ellipticity of the core and the accuracy is low. A major disadvantage of this method is that variations of refractive index along the optical fibre are not known. This method also takes account of leaky modes which are slow to disappear and modify the received energy density.

3.2. The refracted near field method

3.2.1. The principle

To avoid the problem of leaky modes, it is suggested that the light which escapes from the optical fibre at the light input and not the output is examined. A refractive index matching fluid is used to obtain maximum gain.

The convergence angle of the input beam must be greater than the numerical aperture of the optical fibre. The cone of light escaping from the optical fibre is limited by a screen which eliminates the leaky modes and defines the smallest θ''_{min}. The receiver receives the light between θ''_{min} and θ''_{max}; the received power is then given by

$$dP = I_0 \cdot \cos(\theta') \cdot \sin(\theta') \cdot d\theta' \, d\psi \qquad (8)$$

$$P_d(r) = \frac{I_0}{2} \cdot \int_0^{2\pi} d\psi \cdot \int_{\sin^2(\theta'_{min})}^{\sin^2(\theta'_{max})} d(\sin^2(\theta')) \qquad (9)$$

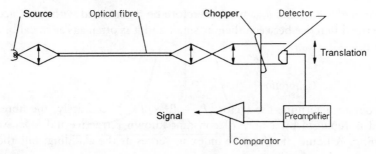

Figure 5.4 Configuration of the near field method

as a function of the minimum and maximum angles of the beam incident on the optical fibre.

By combining Equations (1), (2) and (3) of the previous section, one obtains

$$n_2^2 \cdot \sin^2(\theta') = n^2(r) - n_2^2 + n_2^2 \cdot \sin^2(\theta'') \qquad (10)$$

from which

$$P_d(r) = \pi I_0 \left[\sin^2(\theta'_{max}) - \sin^2(\theta''_{min}) - \frac{n^2(r) - n_2^2}{n_2^2} \right] \qquad (11)$$

which defines the power received by the detector as a function of the maximum angle of incidence and the minimum angle of the beam on the screen.

When the incident beam is directed on to the cladding of the optical fibre, the light detected no longer depends on r:

$$P_{cl} = \pi I_0 [\sin^2(\theta'_{max}) - \sin^2(\theta''_{min})] \qquad (12)$$

by combining Equations (11) and (12), one obtains

$$n^2(r) - n_2^2 = n_2^2 \left[\sin^2(\theta'_{max}) - \sin^2(\theta''_{min}) \cdot \frac{P_{cl} - P_d(r)}{P_{cl}} \right] \qquad (13)$$

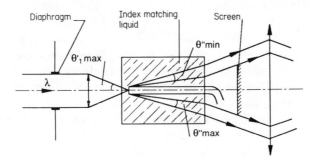

Figure 5.5 Principle of the refracted near field method

The angles θ'_{max} and θ''_{min} must therefore be determined with great accuracy since the difference between them is small and it is often easier to calibrate the system.

3.2.2. Calibration

In order to determine the factor $(P_{cl} - P_d(r))/P_{cl}$ accurately, the fibre to be tested is replaced by a fibre of constant known refractive index n_p without cladding. A liquid of refractive index n_2 serves as the cladding; one then has for the reference fibre

$$n_p^2 - n_2^2 = K \cdot \frac{P'_{cl} - P'_d(r)}{P'_{cl}} \tag{14}$$

for the fibre tested:

$$n^2(r) - n_2^2 = K \cdot \frac{P_{cl} - P_d(r)}{P_{cl}} \tag{15}$$

If a step index reference fibre is used, it can be verified as a first step that the measured refractive index n_p is constant over the whole diameter of the fibre; this enables the quality of the fibre to be confirmed.

3.2.3. Diameter of the opaque screen

It is important to determine correctly the dimensions of the screen which defines the value θ''_{min}. Using the results of Chapter 2 and assuming that Δ is small, it can be shown that

$$\tan(\theta_0) = (2\Delta)^{1/2} \tag{16}$$

It is thus suitable to arrange the screen so that $\theta''_{min} > \theta_0$. Consider, for example, an optical fibre of maximum refractive index $n(r) = 1.467$ in a medium of index 1.455; then $\Delta = 0.014$ and hence $\theta_0 = 9.5°$.

3.2.4. Description of the equipment

A basic diagram is given in Figure 5.6. The electronically controlled motor ensures that the optical fibre is scanned in one direction.

A laser is used since its coherent light can be focused on to a very small surface after being circularly polarized by a quarter wave plate.

The beam is then focused through a 50 μm pinhole to eliminate higher-order transverse modes from the laser; in this way, better focusing on the input face of the optical fibre is obtained.

The $\times 20$ microscope objective has a numerical aperture (of the order of

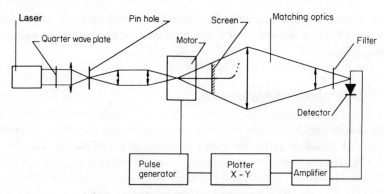

Figure 5.6 Equipment for the near field method

0.5) greater than that of the optical fibre in order to be sure that all the modes are activated and the whole of the angular propagation range of the fibre is covered.

The optical fibre is mounted in a cell filled with index matching liquid whose refractive index is slightly greater than that of the cladding. The cell has two windows: the input window which is formed from a microscope cover slip and the output window. The optical fibre is inserted into the latter, which allows insertion of a hypodermic syringe also containing index matching liquid, so that the fibre can be positioned very precisely at the focal point of the microscope. The cell assembly is mounted on a stage which is controlled by a micrometer screw driven by a synchronous motor so that the optical fibre can be displaced in the laser beam.

By means of two lenses, an image of the pinhole is formed on the detector after passing through a filter set to $0.6328\ \mu m$, which thus allows only the laser light to pass. The opaque screen is suspended by three glass fibres and can be displaced axially. After amplification, the signal is fed to an $X-Y$ plotter which is also coupled to the motor displacement.

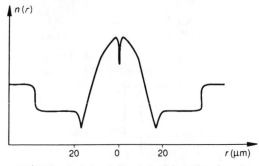

Figure 5.7 Refractive index profile obtained from the refracted near field

The figure shows an example of the result obtained; this method enables an accuracy of $\Delta n \approx 0.0001$ to be obtained and is applicable to monomode optical fibres.

3.3. Interferometric methods

These are in principle more accurate but require the use of an interference microscope which is generally expensive; only two will be described.

The interference microscope actually consists of two microscopes which must be absolutely identical (Figure 5.8). If two identical objects are examined, the field of observation will be uniformly bright or dark according to the relative phase of the two recombined waves. To make observation easier, the object and the reference are inclined with respect to each other and this creates an interference field.

3.3.1. The thin section method

This method consists of observing the interference pattern produced by a thin section consisting of a cross section of the optical fibre. This section must have

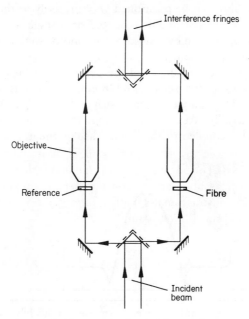

Figure 5.8 The principle of an interference microscope

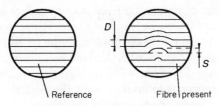

Figure 5.9 Interference fringes

exactly parallel faces and be well polished with a thickness less than the diameter of the optical fibre.

This section will disturb the interference pattern according to the presence of the cladding or a refractive index profile of amplitude S.

Let $\psi = nkd$ be the phase shift of the wave on passing through the thin section of refractive index $n(r)$. The displacement S of a fringe from its initial undisturbed position is associated with a relative change of phase caused by a difference of refractive index between the sample and the reference which can be the cladding:

$$\Delta\psi = (n(r) - n^2)kd \tag{17}$$

from which one obtains:

$$\frac{k(n(r) - n^2)}{S(r)} \cdot d = \frac{2\pi}{D} \tag{18}$$

The inter-fringe distance D and the radial dependence of the fringe deformations $S(r)$ can be measured with a micrometer eyepiece. The fibre and the reference must have the same thickness so they are polished together. The main source of error is the thickness of the section of optical fibre.

3.3.2. The transverse interference method

The section is replaced by the entire optical fibre which is examined transversely to its axis. The advantage of this method is that it is not necessary to prepare the optical fibre apart from immersion in a refractive-index matching liquid. In constrast, mathematical interpretation of the interference fringe deformation is much more complex.

4. MEASUREMENT OF LOSSES IN AN OPTICAL FIBRE

The losses in an optical fibre are of two types, absorption losses and dispersion losses. In both cases full knowledge of the conditions under which light is injected into the fibre is required.

4.1. Injection conditions of light into an optical fibre

A source of light tends to excite the guided modes of an optical fibre in a way which varies according to the mode. A plane incoherent source of equal brilliance for all incidence directions will excite all the modes in the same way on condition that the source is in contact with the optical fibre and larger than it. As the light propagates in the optical fibre, it redistributes itself between the guided modes; this redistribution alters the power carried by each mode.

Without mode coupling and for an infinitely long optical fibre, only the lowest-order mode carries a significant amount of power since the higher-order modes have lost power in proportion to their propagation distance. Mode coupling will therefore optimize the power carried by the different modes and an equilibrium state will be established. Consequently, the power carried by a mode with respect to a given reference mode no longer depends on the length of the fibre. Once this equilibrium state is reached, the total power transmitted obeys a law of exponential decrease with a unique loss coefficient instead of having a loss coefficient for each mode. It is this coefficient which enables the loss per unit length of an optical fibre to be defined.

The injection conditions for an optical fibre will be chosen in order to create an equilibrium state as rapidly as possible. Typical light injection into an optical fibre is shown in Figure 5.10.

The source can be a laser or an incoherent light. Two parameters define the beam — the spot diameter and its numerical aperture. The two parameters must suit the optical fibre. The spot diameter must be equal to the core diameter of the fibre and the numerical aperture of the spot must be equal to that of the optical fibre; that is, the maximum angle of the modes which can be guided in the fibre (see chapter 2)

$$NA = \frac{\left(n^2(r) - n_2^2\right)^{1/2}}{n_0} \tag{19}$$

For a graded index optical fibre, the numerical aperture is maximum for the maximum value of refractive index and varies with the radius of the optical fibre; it is, therefore, difficult to design a light source which satisfies these conditions.

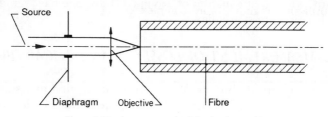

Figure 5.10 Arrangement for injection into a fibre

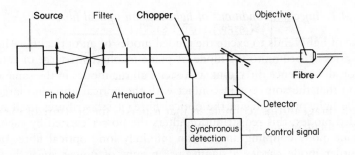

Figure 5.11   Conventional arrangement for injection of light into an optical fibre

A typical system for injecting a beam into an optical fibre to perform measurements is shown in Figure 5.11.

The source can be a laser, a lamp or a light-emitting diode. This light is subjected to spatial filtering by passing it through a pinhole (50 μm) and spectral filtering by passing it through a filter set to λ_0 (an interference filter or monochromator). An attenuator is used so that the power remains constant in spite of fluctuations of the source and for any wavelength.

A modulator and associated beam splitter are used to perform measurement by synchronous detection.

Finally the light passes through a microscope objective containing an aperture which optimizes coupling with the optical fibre. Alignment of the optical fibre and the objective is achieved by a three-axis displacement system and, if necessary, a system which permits the influence of the cladding to be avoided.

4.2. Mode scramblers

Mode scramblers enable an equilibrium state of modes to be obtained more quickly than in a fibre on its own.

If an optical fibre has significant mode coupling, it can be used as a mode scrambler before being coupled to the fibre to be measured.

Another standard scrambler consists of three optical fibres, each of one metre length, with an attenuation of around 1.5 dB; one has a graded index and the other two are step index.

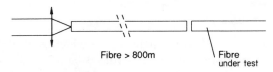

Fibre > 800m   Fibre under test

Figure 5.12   Use of a long fibre as a mode scrambler

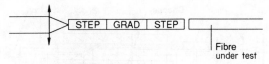

Figure 5.13   A mode scrambler

The advantage of this scrambler is that it makes the aperture of the output ray independent of the injection conditions.

4.3. Total attenuation per unit length

One of the important parameters in the design of optical fibre links is the total loss as a function of length, indendently of the propagating modes. This parameter defines the loss suffered by an injected luminous flux as a function of the distance travelled in the optical fibre.

The measurement is performed, in principle, by using a source for which the power injected into the fibre is known, a mode scrambler, the fibre under test and a detector. Knowledge of the power injected into the optical fibre, and that emerging from it, enables the loss per unit length, expressed in dB/km, to be defined. If the losses are assumed to be of exponential form, one has:

$$A\,(\mathrm{dB/km}) = \frac{10 \cdot \log[P(L) - P(0)]}{L} \tag{20}$$

where L is expressed in km, $P(L)$ is the power received by the detector and $P(0)$ is the injected power.

4.4. Measurement of absorption losses

Absorption losses can be measured only by calorimetry although this method lacks accuracy and introduces systematic errors.

Absorption in a material leads to heating, which can be measured. Although simple in principle, this method is very difficult to use since the temperature variation is very small.

4.4.1. The principle

The optical fibre is excited with a light source which is of high intensity but is insufficient to cause non-linear effects (the Brillouin and Raman effects).

The calorimeter includes a thermal screen in order to reduce temperature

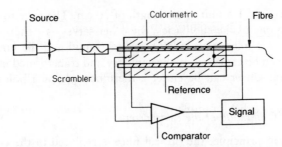

Figure 5.14 The principle of loss measurement by absorption

fluctuations; it is sufficiently long for the heat from the optical fibre supports to be negligible.

If the light source is turned on at time $t = 0$, the optical fibre starts to warm by absorption. The quantity of heat Q produced per unit time and length is proportional to the injected power P.

$$Q = 2\alpha_a \cdot P \tag{21}$$

where α_a defines the loss of power attributable to absorption.

As soon as the temperature of the optical fibre exceeds the equilibrium temperature, heat flow proportional to the temperature difference occurs in accordance with the following equation:

$$q = C_1(T - T_0) \tag{22}$$

where C_1 defines the combined conduction and radiation losses.

The temperature of the optical fibre increases due to the heat Q generated in the fibre and decreases due to the heat q lost to the surrounding medium; this can be written as follows:

$$K \cdot \frac{dT}{dt} = Q - q \tag{23}$$

where K is the thermal capacity per unit length of the optical fibre. At time $t = 0$ and hence $q = 0$, one has

$$2\alpha_a = \frac{K}{P} \cdot \left(\frac{dT}{dt}\right)_{t=0} \tag{24}$$

Measurement of the loss at the initial time together with knowledge of P and K enable the absorption parameter α_a to be determined.

4.4.2. Description of the calorimeter

The external enclosure, consisting of a copper tube, contains two identical parallel capillaries in fused silica with an internal diameter of 0.5 mm, an external diameter of 1.5 mm and a length of 20 cm. The fibre to be measured is inserted into one of the capillaries; the other serves as a thermal reference. The capillaries are filled with index matching liquid and the power escaping from the fibre can be captured by the capillary and transformed into heat. The temperature transducer can be either a thermocouple or a bolometer.

4.4.3. Calibration of the calorimeter

This is simple in principle; the optical fibre is replaced in the capillary by a fine electric wire of known diameter. This serves to calibrate the calorimeter by means of a known electric current which causes heating and can be accurately measured. The thermal capacity is not exactly the same when the capillary contains the optical fibre instead of the wire; this error is generally negligible, however.

4.4.4. Disadvantages of this method

This method has numerous disadvantages which limit its usefulness:

— A very small temperature variation must be measured with great accuracy.
— Calibration introduces a systematic error.
— Long-term temperature drift is difficult to evaluate.
— Scattered energy absorbed by the temperature sensor is difficult to determine.
— An additional source of error arises from leaky modes in the cladding which can penetrate into the calorimeter.

4.5. Losses due to scattering

Scattering losses result from two phenomena: Rayleigh scattering which varies as $1/\lambda^4$ and scattering due to defects in the optical fibre.

The aim in this case is to measure the light scattered by the optical fibre in order to determine the scattering coefficient $2\alpha_s$. It is known that

$$P(z) = P(0) \cdot \exp(-2\alpha_s \cdot z) \qquad (25)$$

This coefficient can thus be obtained from the power ΔP_z scattered out of the fibre in a given length D for a total transmitted power $P(z)$ using the following expression:

$$2\alpha_s = \frac{\Delta P(z)}{D \cdot P(z)} \qquad (26)$$

To measure this quantity, an integrating sphere is used; the interior is painted white so that radiation scattered by the optical fibre is not absorbed but reflected to the detector whatever the angle of emission of the scattered light.

No light must reach the detector directly, so a baffle is used. The detector operates in synchronous mode in order to capture the light scattered by the fibre when it transmits a modulated flux.

One of the advantages of this method is that it enables fluctuations and losses along the fibre to be measured by displacing it in the sphere.

The accuracy of this method depends on the quality of the integrating sphere, the choice of reflecting paint for the wavelength used and the accuracy with which $\Delta P(z)$ and $P(z)$ can be measured. It is preferable to measure these two parameters with the same detector; this is possible by routeing the output end of the optical fibre into the sphere. It is, however, necessary for the detector to have a linear response for both the low power of scattering losses and the high power of the flux in the optical fibre.

4.6. Measurement of spectral attenuation

It is known that spectral attenuation plays a dominant role in the losses of monomode optical fibres and it must be measured to determine the loss factor of the fibre. The principle is to measure the luminous flux emerging from the fibre as a function of the injected wavelength. To eliminate the system transfer function, two measurements are performed, one with a long optical fibre (≈ 3 km) and the other with a short fibre (a few metres); the loss factor can then be written:

$$\alpha(\lambda) = \frac{10}{L - L_1} \cdot \log\left[I_L(\lambda) - I_1(\lambda)\right] \tag{27}$$

The system consists of a monochromator delivering a constant power for all measurements, optics for injection into the fibre, a modulator and a detector with wide dynamic range to provide synchronous detection.

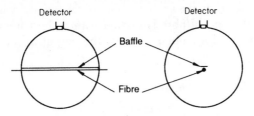

Figure 5.15   Configuration of the integrating sphere

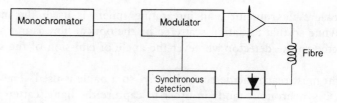

Figure 5.16   The principle of spectral attenuation measurement of a fibre

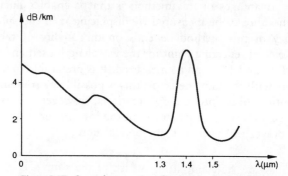

Figure 5.17   Spectral attenuation of a monomode optical fibre

With this type of system, a wavelength resolution of 5 nm and an accuracy of 0.5% in the attenuation factor can be obtained.

It should be noted that, for $d = 1.3$ μm and 1.55 μm, the attenuations are almost identical and of the order of 0.5 dB/km. The critical points in this arrangement are the optical interfaces, which must be perfectly corrected for chromatic abberation at 0.5 μm and 1.6 μm, and preparation of the fibre ends in order to limit insertion losses and disparities between measurements on the same optical fibre.

5. MEASUREMENT OF THE FREQUENCY CHARACTERISTICS OF OPTICAL FIBRES

Frequency response measurements can be performed in the frequency domain or in the time domain. In the first case, the source frequency is modulated and, in the second, the optical fibre is excited with a short luminous pulse. The signals to be analysed are obtained from the output of the optical fibre.

5.1. Definition of the source

Let $S(\lambda)$ be the profile of the source, centred on λ_0, which is assumed to be of Gaussian form:

$$S(\lambda) = S(\lambda_0) \cdot \exp\left[-\frac{1}{2} \cdot \left(\frac{\lambda - \lambda_0}{\sigma_s} \right)^2 \right] \tag{28}$$

such that

$$\int_0^\infty S(\lambda) \, d\lambda = 1$$

the variance of this source (or RMS value) can be written

$$\sigma_s = \left[\int_0^\infty (\lambda - \lambda_0)^2 S(\lambda) \, d\lambda \right]^{1/2} \tag{29}$$

One has

$$\sigma_s = 0.424\Delta\lambda_{1/2} \qquad \text{at half amplitude} \tag{30}$$

$$\sigma_s = \frac{\Delta\lambda_{1/e}}{(8)^{1/2}} \qquad \text{at } 1/e \text{ of the amplitude} \tag{31}$$

which determines the 'half amplitude' width of the source.

5.2. The transfer function

The output signal of an optical fibre $P_2(t)$ is the convolution of the input signal $P_1(t)$ and the transfer function $h_g(t)$ of the optical fibre:

$$P_2(t) = \int_{-\infty}^{+\infty} h_g(t - \tau)P_1(\tau) \, d\tau \tag{32}$$

which, in the Fourier domain, can be written

$$H_2(\omega) = H_g(\omega) \cdot H_1(\omega) \tag{33}$$

where

$$H_i(\omega) = \int_{-\infty}^{+\infty} P_i(t) \cdot \exp(-j\omega t) \, dt \tag{34}$$

is the Fourier transform of $P_i(t)$.

5.3. Measurement of pulse spreading

If a Gaussian impulse $P_1(t)$ is fed into an optical fibre, a slightly different Gaussian distribution $P_2(t)$ is obtained at the output. Deconvolution of the

output gives the frequency spectrum of the optical fibre:
at the input:

$$P_1(t) = P_1 \cdot \exp\left(-\frac{t^2}{2\sigma_1^2}\right) \tag{35}$$

at the output;

$$P_2(t) = P_2 \cdot \exp\left(-\frac{t^2}{2\sigma_2^2}\right) \tag{36}$$

hence, in the Fourier domain,

$$H_f(\omega) = P_1(\omega) \cdot P_2(\omega) \tag{37}$$

which, after normalization, can be written:

$$H_f(\omega) = \exp\left[-\frac{\omega^2(\sigma_2^2 - \sigma_1^2)}{2}\right] \tag{38}$$

where $\sigma_t^2 = \sigma_2^2 - \sigma_1^2$ is the sum of the intermodal and intramodal dispersions; it enables the extent of pulse spreading to be determined.

5.4. Definition of the cut-off frequency

The cut-off frequency determines the transmission capacity of the optical fibre. Measurement of the frequency response of optical fibres is performed by using detectors which generate photocurrents proportional to the received power.

$$i_{ph} = S \cdot P_2(t) \tag{39}$$

where S is the sensitivity of the detector in A/W. The electrical power which is developed in the detector load resistance R_L can be written

$$P_e(t) = R_L \cdot i_{ph}^2 = R_L S^2 P_2^2(t) \tag{40}$$

which can be written in the frequency domain

$$|H(\omega)|_{dB} = 10 \cdot \log(H_e(\omega)) = 20 \cdot \log(H_2(\omega)) \tag{41}$$

Hence the cut-off frequency at -3 dB of the optical characteristic corresponds to a cut-off frequency at -6 dB of the processed electrical signal. Using the previous relations, the cut-off frequency at -6 dB can be written

$$F_{-6dB} = \frac{0.187}{\sigma_t} = \frac{0.44}{[(\Delta t_{1/2})_2^2 - (\Delta t_{1/2})_1^2]^{1/2}} \tag{42}$$

where $\Delta t_{1/2}$ defines the half-amplitude width in absolute time space.

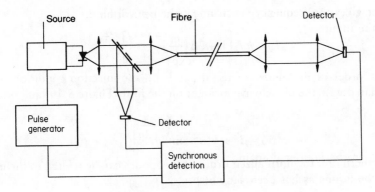

Figure 5.18 Measurement of the frequency response of an optical fibre

5.5. Equipment

The principle is to measure the spreading of the output pulse by comparing it with the input pulse (Figure 5.18).

The optical beam, modulated by the pulses, is separated into two beams, one towards the input detector and the other towards the optical fibre.

The signals from the input and output detectors are sent to an oscilloscope which is triggered by a signal provided by the pulse generator.

The performance of this method is poor since it is affected by the conditions of injection into the fibre and reception at the detector. Furthermore, the pulse method is not the most efficient in respect of signal-to-noise ratio at the output of the optical fibre. It is preferable to use a baseband modulation method which has a 50 times better signal-to-noise ratio. Nevertheless, it has the advantage of being simple and giving a good approximation to the frequency response.

Its main advantage is the measurement of pulse spreading whatever the cause of this spreading (intermodal or intramodal dispersion).

6. OBSERVATION OF MODE GROUPS

6.1. The principle of measurement

This type of measurement is applicable only to multimode optical fibres as shown by the relations established in Chapter 2. Recall that the propagation constant $\beta_{\mu f}$ of the mode $m = 2\mu + |f|$ can be written

$$\beta_{\mu f} = kn_1 \left[1 - 2\Delta \left(\frac{m}{M}\right)^{2\alpha/\alpha+2} \right]^{1/2} \tag{43}$$

The order of the mode excited depends on the injection conditions (r, θ) associated with the plane wave incident on the fibre (Figure 5.19) and one has

$$k^2 n^2(r) = \beta^2 + k^2 r + k^2 \psi \tag{44}$$

$$k^2 n^2(r) = \beta^2 + k^2 n^2 \sin^2(\theta_1)\sin^2(\phi) + f^2/r \tag{45}$$

Furthermore, the normalized order of the modes can be related to the injection conditions as has been shown:

$$\left(\frac{m}{M}\right)^{2\alpha/(\alpha+2)} = \left(\frac{r}{a}\right)^a + \left(\frac{\sin(\theta_e)}{\sin(\theta_c)}\right)^2 \tag{46}$$

where θ_e is the angle of injection and θ_c is the numerical aperture.

Observe that the azimuthal order f of a mode is related to the angle θ_e and the radial position r; hence variation of the angle for excitation conditions centred on $r = 0$ enables modes characterized by $f = 0$ to be excited:

$$\frac{m}{M} = \left(\frac{\sin(\theta_e)}{\sin(\theta_c)}\right)^{(\alpha+2)/\alpha} \tag{47}$$

and similarly if $\theta_e = 0$, that is with incidence parallel to the optical axis:

$$\frac{m}{M} = \left(\frac{r}{a}\right)^{(\alpha+2)/2} \tag{48}$$

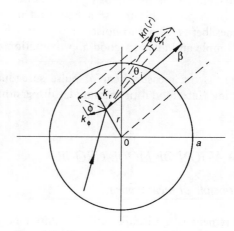

Figure 5.19   The various components of a wave vector

For a given excitation beam, the spread in the propagation constant β can be written

$$\delta B = kn_1\Delta\left[\frac{4r_0\delta r}{a^2} + \left(\frac{\delta\theta}{\theta_c}\right)^2\right] \quad \text{where } \beta \approx kn_1[1 - \Delta R^2] \quad (49)$$

Use of a monomode optical fibre enables optimum excitation conditions to be realized with the condition that the ratio $d\beta/dr$ is minimized.

If the beam emerging from the optical fibre is considered to be Gaussian, which is a good approximation for a monomode optical fibre, one has

$$P(r) = \exp\left[-2\left(\frac{r - r_0}{w_0}\right)^2\right] \quad (50)$$

$$P(\theta) = \exp\left[-2\left(\frac{\theta}{\delta\theta}\right)^2\right] \quad \text{where } \delta\theta = \frac{\lambda}{\pi w_0} \quad (51)$$

where w_0 is a constant of the Gaussian distribution defining the width of the excitation spot; in this case one has

$$\delta\beta = kn\Delta_1\left[\frac{4r_0w_0}{a^2} + \frac{\lambda^2}{\pi^2 w_0^2\theta_c^2}\right] \quad (52)$$

Minimization of $\delta\beta$ implies a value of w_0 such that

$$w_0 = \left[\left(\frac{\lambda a}{\pi\theta_c}\right)^2 \frac{1}{2r_0}\right]^{1/3} \quad (53)$$

which signifies that for each radial value r_0 of excitation of the fibre, there is an optimum value w_0 which enables the number of excited modes to be minimized and hence selective excitation to be increased.

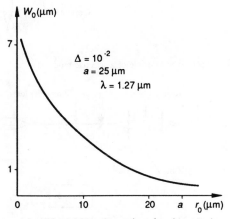

Figure 5.20 Excitation spot size which minimizes the number of modes excited as a function of radial position

6.2. Experimental configuration

6.2.1. Selective intermodal analysis

In this configuration (Figure 5.21), a small group of modes is excited by using a single-mode optical fibre for the source wavelength and exploiting the variations in propagation time with the order of the modes excited.

6.2.2. Selective modal excitation

In this configuration (Figure 5.22), all the modes which can propagate in the fibre are uniformly excited; analysis is performed with the help of a polar diagram at the output of the optical fibre and phase variations are measured.

6.3. Summary of methods

Use of these methods enables various information concerning the characteristics of optical fibres to be obtained:

— the variation of group delay time as a function of λ;
— the influence of mode coupling;
— the magnitude of intermodal dispersion;
— the mode amplitude.

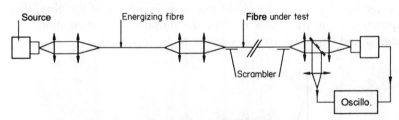

Figure 5.21   The principle of a selective mode analyser

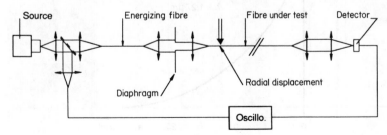

Figure 5.22   The principle of selective excitation

These parameters enable the deviation of the refractive index from the optimum and the frequency response to be determined. This method can also be used with monomode optical fibres.

7. BACKSCATTERING METHODS

7.1. Introduction

Backscattering is one of the most used methods of measuring attenuation both by manufacturers of optical fibres and by installers. It enables non-destructive inspection of fibre optic links to be performed by measuring the attenuation of a fibre, cracks to be detected and the influence of connections to be determined.

7.2. The theory of backscattering

Let $E(z)$ be a luminous pulse injected at $z = 0$ and time $t = 0$ into an optical fibre:

$$E(z) = E_0 \exp\left[- \int_0^z \alpha'(l)\, dl\right] \tag{54}$$

where $\alpha'(l)$ defines the absorption coefficient as a function of the wavelength propagating in the fibre. It is assumed that all modes have the same attenuation and the integral represents the cumulative attenuation over a distance z.

Over an interval $[z, z + dz]$ part of the light wave is scattered in a solid angle of 4π with a scattering coefficient $\alpha_d(z)$:

$$dE(z) = \alpha_d(z) \cdot E_0 \cdot \exp\left[- \int_0^z \alpha'(l)\, dl\right] dz \tag{55}$$

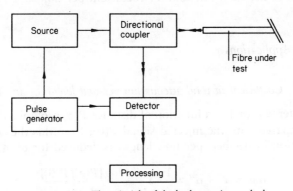

Figure 5.23 The principle of the backscattering method

Let $S(z)$ be the fraction of energy scattered backwards in the acceptance angle of the optical fibre. The quantity of backscattered energy which returns to the source is

$$dE(z) = S(z) \cdot \alpha_d(z) \cdot E_0 \cdot \exp\left[- \int_0^z \alpha'(l)\, dl \right] dz \qquad (56)$$

This return occurs with an attenuation factor $\alpha''(l)$ such that the energy at the input of the optical fibre due to scattering in the interval $[z, z + dz]$ is

$$dE = S(z) \cdot \alpha_d(z) \cdot E_0 \cdot \exp\left[- \int_0^z (\alpha'(l) + \alpha''(l))\, dl \right] dz \qquad (57)$$

$S(z)$ is called the backscatter factor. Let v_g be the group velocity of the luminous energy; the round trip time is

$$t = 2z/v_g$$

from which

$$dt = 2\, dz/v_g$$

The expression for the power $P(t)$ measured by backscattering for an input pulse width τ can be written

$$P(t) = 0.5 P_0 \tau S(z(t)) \alpha_d(z(t)) v_g \exp\left[- \int_0^{z(t)} (\alpha'(l) + \alpha''(l)\, dl \right] \qquad (58)$$

if P_0 is the injected power. It can be shown that, for a graded index optical fibre, an expression for the backscatter factor is

$$S(z) = \frac{3}{8} \cdot \frac{n^2(z,0) - n^2(z,a)}{n^2(z,0)} \cdot \frac{\alpha(z)}{\alpha(z) + 1} \qquad (59)$$

and $\alpha(z)$ defines the total attenuation coefficient per unit length.

7.3. Applications

7.3.1. Coefficient of total attenuation per unit length of an optical fibre

The backscattered signal as a function of distance $P(z)$ is obtained experimentally and compared with the injected signal $P(0)$; from this the coefficient of total attenuation in decibels per unit length is deduced for $\alpha' \leqslant \alpha''$:

$$\alpha_{dB} = \alpha' + \alpha'' = \frac{1}{2} \cdot \Delta\, \frac{10\, \log[P(z)/P(0)]}{\Delta z} \qquad (60)$$

7.3.2. Crack detection

The signal reflected by a crack of reflection factor R is

$$P_c(z) = RP_0 \cdot \exp\left[- \int_0^z \left(\alpha'(z) + \alpha''(l) \right) \, dl \right] \qquad (61)$$

where z is the location of the crack. Let $P(z)$ be the signal just before the crack, one has

$$\frac{P_c(z)}{P_r(z)} = \frac{R}{0.5\tau\alpha_d(z)S(z)v_g} \qquad (62)$$

and the losses introduced by the crack can be written

$$L_{\text{cra}} = 10 \, \log\left(\frac{P_c(z)}{P_r(z)}\right) \qquad (63)$$

7.3.3. Losses due to splices and connectors

The backscattering method does not enable the actual signal losses due to a splice or a connector to be measured. This impossibility is due to the fact that the backscattered light follows a round trip path in the fibre; the measurement relates to insertion losses corresponding to a double passage and hence does not give access to the losses for a single passage as is the case for connectors and splices.

7.4. Example of backscattering measurement

Consider an optical fibre where absorption is neglected; the backscattered power can then be written

$$P_r(z) = 0.5 P_0 \tau S \alpha v_g \exp(-2\alpha z)$$

with the following numerical values:

$$\lambda = 0.8 \ \mu\text{m} \qquad NA = 0.18 \qquad n_1 = 1.46 \qquad \tau = 100 \ \mu\text{s}$$

the backscattering factor is deduced as $S = 3.8 \times 10^{-3}$.

Measurement gives $2\alpha_1 = 10$ dB for an optical fibre of length 2 km, hence

$$\alpha = \alpha_d = 2.5 \ \text{dB}$$

The group velocity can be written

$$v_g = c/n_1 = 2.05 \times 10^{+8} \ \text{m/s}$$

From which one deduces

$$P_r(z) = P_0 \times 2.3 \cdot 10^{-5} \cdot \exp(-2\alpha z)$$

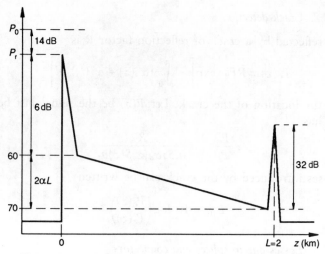

Figure 5.24   An example of measurement of the backscattered signal

and

$$-10 \log\left[\frac{P_r(z)}{P_0}\right] = 46.3 + 2\alpha z$$

Observe that the initial value of backscattering is 46.3 dB. If an output crack is now considered with a reflection coefficient of 4%, one obtains

$$\frac{P_c}{P_r} = 1.7 \times 10^{+3}$$

hence a loss due to the crack of:

$$L_{cra} = 32.3 \text{ dB}$$

This type of measurement is performed every day by cable installers since it enables them to observe the state of the fibre after its installation and to verify the absence of faults.

8. MEASUREMENTS SPECIFIC TO MONOMODE OPTICAL FIBRES

Two measurements are performed particularly on monomode optical fibres; these are measurement of the cut-off wavelength and the mode width.

8.1. Measurement of the cut-off wavelength

Be definition, the cut-off wavelength is that above which the HE_{11} mode is the only one which propagates. In practice, it is necessary to take account of the actual conditions of use of the fibre tested and its radius of curvature.

Below a certain radius of curvature, the losses increase and this greatly increases absorption by the fibre for a given wavelength.

The losses are due to increasing coupling between guided modes and radiating modes. The first absorption peak corresponds to extinction of modes of order higher than HE_{10}.

For measurement, the same technique as for spectral attenuation is used but with a short optical fibre having a radius of curvature between 5 and 50 mm.

8.2. Measurement of the mode width

There are numerous methods which use the Gaussian field approximation; that of the Ronchi mask which can be adapted to spectral measurement is notable.

The principle of this method is to use a mask which consists of two absorbent bands and one transparent one whose width is proporational to the half amplitude width of the Gaussian beam. The light beam emerging from the optical fibre passes through this aperture plate and is focused on to a detector.

The mask is located at a distance d from the optical fibre and its displacement enables positions of minimum (P_m) and maximum (P_M) intensity to be obtained.

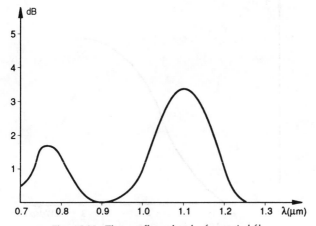

Figure 5.25 The cut-off wavelength of an optical fibre

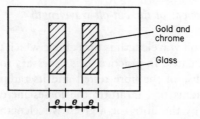

Figure 5.26 Configuration of a Ronchi mask

The first step in this method is to establish a chart which relates the received intensity to the width of the aperture; the expression for the intensity is

$$I(r, d) = I_0 \cdot \exp\left(\frac{r^2}{w_d^2}\right) \tag{64}$$

where I is a function which depends on the distance from the fibre and is defined for a width w_d at $1/e$.

It can be shown that the width of the beam w_d is inversely proportional to the width of the propagating mode:

$$w_0 \approx \frac{d\lambda}{\pi w_d} \tag{65}$$

The limitation of this method is the limitation of the Gaussian model used to represent the field. Use of this system with a monochromator, of the type used for spectral attenuation, enables the variation of the mode width as a function of wavelength to be obtained.

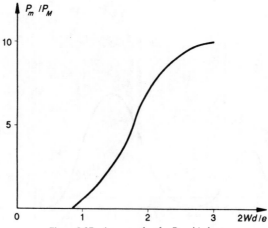

Figure 5.27 An example of a Ronchi chart

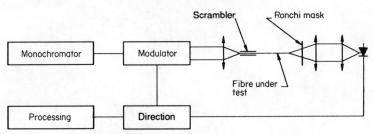

Figure 5.28   Configuration for measuring mode width

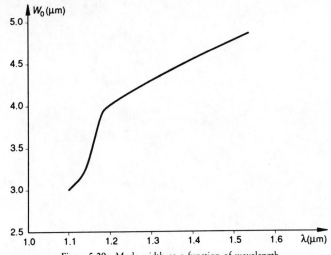

Figure 5.29   Mode width as a function of wavelength

As with all methods of this type, a germanium detector is used with synchronous detection to obtain a wide dynamic range.

Figure 5.29 shows an example of a curve representing the mode width as a function of wavelength obtained by this method.

The considerable advantage of these two methods is that they can be included in the same measurement system used for spectral attenuation; this limits the required manipulation of the fibre and the hardware.

6

FIBRE OPTIC CABLES

1. INTRODUCTION

Two aspects must be considered in the design of fibre optic cables. One is the poor mechanical properties of the fibre, due to its small diameter (a few hundred microns) and great length (1–3 km in one section). The other is the high sensitivity to strain caused by bending radiuses which cause microfractures; these lead to transmission losses which can amount to several dB/km.

Observe that these losses become very substantial for a radius of curvature less than 60 mm.

The application of strain to an optical fibre induces additional attenuation which is particularly significant if the strain causes deformation of the fibre axis. The deformation appears to the light wave as a variation of longitudinal refractive index which causes either mode coupling in the case of curvatures having intervals of the same order as the optical path (see Chapter 2) or a loss of high-order modes for curvatures of sustained radius.

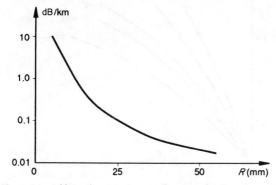

Figure 6.1   Additional attenuation as a function of radius of curvature

2. MECHANICAL PROPERTIES AND ENDURANCE OF OPTICAL FIBRES

The reliability of an optical fibre link depends on its mechanical strength and its endurance.

2.1. Tensile strength

The theoretical tensile strength of a thread of silica 125 μm diameter is 250 N but the values presently obtained (20–50 N) are explained by the presence of surface cracks which concentrate the strains. Present techniques enable homogeneous distribution of faults to be obtained and tend to limit extrinsic faults.

2.2. Fatigue

Static fatigue of the glass, caused by ambient humidity, results in an increase of surface cracking when a charge is applied. This increase also depends on the temperature and it increases rapidly as the temperature increases.

Observe that in Figure 6.2 when the service stress τ_c is less than 20% of the

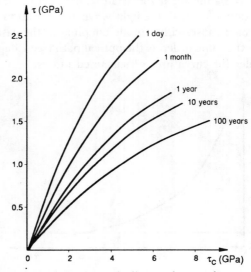

Figure 6.2   Probable lifetime of a fibre as a function of service stress

Table 6.1 Primary protection of various optical fibres

O.F.	Cladding (μm)	Silicone (μm)	Hytrel (μm)
50/125	125	260	550
100/140	140	260	550
200/280	280	450	650

initial stress τ_a, the effect of static fatigue is negligible; in contrast, if the service stress τ_c is greater than 30% of the initial stress τ_a, the lifetime does not exceed a few days.

3. PROTECTION OF OPTICAL FIBRES BEFORE CABLING

It has been seen that during fabrication, the optical fibre is surrounded by a protective layer, generally of silicone or epoxy resin, immediately after the drawing operation. This primary coating to a thickness of several microns protects the optical fibre from external attack (due to humidity and abrasion) and minimizes the role of microfractures. The coating is chosen to have a coefficient of thermal expansion similar to that of the optical fibre in order to limit mechanical stresses due to differential expansion.

This first protection generally consists of two compact sheaths produced by extrusion around the optical fibre during drawing; the first sheath is of silicone on to which hytrel is extruded.

4. MATERIALS USED IN CABLES

In choosing materials, two aspects must be considered. The first is to ensure that the mechanical properties are satisfactory and suitable for an industrial environment. The second is to provide sealing which avoids degradation of the optical fibre and permits its use as an electrical insulator.

4.1. Mechanical reinforcement of cables

Optical fibres cannot support high stretching forces without risk of breakage and so materials are used to reinforce them; these materials have a high Young's modulus and the disadvantage of a high stiffness. The main materials used are steel, glass fibre, synthetic fibres, carbon fibres and boron fibres.

Table 6.2 Mechanical characteristics of carriers

	Density	Breaking strength 10^2 N/mm^2	Young's modulus 10^2 N/mm^2	Breaking strain %	B_{sp}
Steel wire	7.38	30	2000	2	3.83
Kevlar 29	1.44	27	620	4	18.7
Kevlar 49	1.45	27	1300	2	18.6
Carbon fibre	1.95	20	4000	0.5	10.2
Glass fibre	2.54	17	700	2.4	6.69

Steel is the most used material since it has a high Young's modulus ($200 \, \text{kN/mm}^2$), is very strong and costs little; on the other hand, it is an electrical conductor and increases the weight of cables.

Synthetic and organic fibres are also used in the form of braid, strands or filaments. The most widely used is Kevlar and these fibres are also good electrical insulators.

As a parameter for comparison, the specific breaking strength (B_{sp}) is defined by the relation:

$$B_{sp} = \frac{\text{breaking strength}}{\text{relative density}}$$

Table 6.2 enables the various materials used to be compared. Observe that Kevlar and steel have comparable breaking strengths but a factor of 6 in the weights gives the advantage to Kevlar [B_{sp}(Kevlar) $\approx 5B_{sp}$(steel)]. The latter has the disadvantage of very poor characteristics in compression and this limits its use.

4.2. Materials used for sealing

One of the factors affecting the endurance of an optical fibre is humidity, which enlarges surface cracks. That is why an impermeable barrier is introduced into cables; the most used barrier is a metallic one.

One of the most used forms of protection is a tape consisting of a band of aluminium several tens of millimetres thick interleaved with one or two polyethylene tapes. This tape is applied as a covering during the extruding operation of the external plastic sheath. The heat and pressure of the extruded plastic are sufficient to ensure welding of the two edges of the tape, which ensures sealing. Tensile forces are thus supported by the cable covering and not the optical fibre itself.

5. THE STRUCTURE OF OPTICAL CABLES

In connection with fabrication, it is convenient to distinguish two classes of cable — those with a compact structure and those with a loose structure.

In compact structures, the fibre is embedded in a plastic material whose behaviour has an influence on the transmission medium. This technology is highly developed for plastic fibres where the attenuation, before cabling, amounts to hundreds of dB/km.

In loose structures, three classes of cable can be distinguished:

— grooved cylindrical cables;
— conventional cables;
— ribbon cables.

5.1. Optical cables with a conventional structure

The fibre is enclosed in a plastic sheath in which is free from all stresses. Fibres of Kevlar or other material are inserted in this sheath and this provides tensile strength.

The second stage is installation of the impermeable barrier and finally the exterior coating is applied according to the conditions of use — interior, exterior or underground.

Figures 6.3 and 6.4 show examples of conventional single-fibre cables. These fibres, with their PVC coating, can be assembled to form fibre bundles. There is, therefore, an infinite variety of cable configurations.

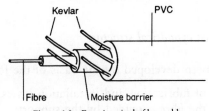

Figure 6.3   Exterior single-fibre cable

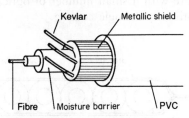

Figure 6.4   Underground single-fibre cable

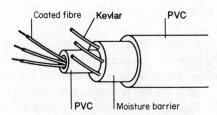

Figure 6.5 Exterior multichannel cable

5.2. Sheet or ribbon structure

Obtaining a good coefficient of filling is one of the criteria which lead to the production of ribbon structures. The fabrication method consists of arranging the fibres on a support in a horizontal plane and then covering them with a matched supporting cover; this can be either loose or compact.

In the case of a compact structure, the support for the sheet of fibres consists of two ribbons joined together by an adhesive material. The loose support consists of two preformed ribbons (plastic or metallic) which are superposed and contain as many grooves as there are fibres. The fibres are deposited without strain in the cavities before welding or bonding the two parts.

A cable is obtained by superposing ribbons and twisting; this is very difficult due to the unequal distribution of strains within the cable. At present, cables realised in this way have increased attenuation due to cabling and more frequent breaks which cannot be repaired.

5.3. Grooved cylindrical structure

This structure has been developed on the basis of the following features:

— total absence of fabrication and installation losses;
— protection of the fibre provided by the support;
— modular structure with a small number of optical fibres;
— use of fibres with only a single coating.

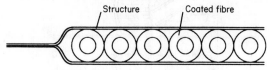

Figure 6.6 Compact ribbon type

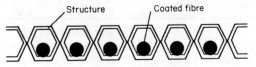

Figure 6.7 Loose ribbon type

The cabling element consists of a cylindrical rod of diameter D in which the fibres of external diameter d are supported without tension and with a slight over-length ε_t. The grooves, of depth h, are helicoidal and continuous and this ensures cabling of the fibre without strain.

The cylindrical former is reinforced with a central carrier which provides the mechanical and thermal properties of the assembly. The behaviour of the cable is determined by the following parameters:

(a) The difference in length (ε_c) between the fibre path at the bottom of the groove and the path at the top of the groove:

$$\varepsilon_c = \frac{2\pi^2(D-h)(h-d)}{p^2 + \pi^2(D-d)^2} \tag{1}$$

(b) The helix of the former imposes a continuous radius of curvature R on the fibre:

$$R = \frac{(D-d)(1+p^2)}{2\pi^2(D-d)^2} \tag{2}$$

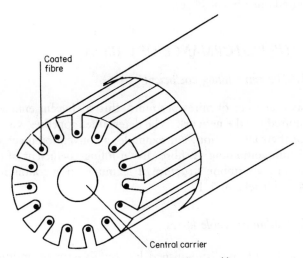

Figure 6.8 Slotted structure for 16 cables

Table 6.3   Mean attenuation of a slotted cable

	Attenuation in dB/km		
	$\lambda = 0.84\ \mu m$	$\lambda = 1.30\ \mu m$	$\lambda = 1.55\ \mu m$
Before cabling	2.7	0.6	0.4
After cabling	2.7	0.5	0.3

(c) When the fibre meanders in the groove, a periodic radius of curvature (R_g) can arise (determined to be 100 mm) which leads to an over-length ε_g:

$$\varepsilon_g = \frac{h - d}{8R_g} \tag{3}$$

(d) The combination of these two curvatures will induce a static bending stress σ_a which must not exceed a certain limit and this depends on the required lifetime of the optical fibre.

The coefficients of expansion and the mechanical behaviour of the material must also be considered. It is, therefore, necessary to compare the geometrical parameters and the mechanical and thermal parameters in order to optimize the qualities of the cable for each application.

Notice that the fibre attenuates less after cabling than before. It is assumed that this variation is due to the disappearances of losses caused by microbending during processing on the fibre drum which disappear during cabling.

Here again, several grooved elements can be combined to form cables of several hundreds of optical fibres.

6. THE PERFORMANCE OF CABLES

6.1. The cable filling coefficient

Expressed as a number of fibres per mm^2, this coefficient enables the size of conduits required for the network to be determined together with the material necessary for their installation. For example, a conventional cable of external diameter 18 mm containing 12 fibres has a filling coefficient of 0.07 while a cable with a compact ribbon structure of 12 mm section containing 144 fibres has a filling coefficient of 1.12.

6.2. Additional cable losses

A good optical cable is distinguished by the absence of additional cabling losses. The previous analysis shows that only cables with a free structure are capable of preserving the mean attenuation of optical fibres.

6.3. Mechanical behaviour

The mechanical behaviour of cables depends on the nature of their reinforcement and their mode of application. For each cable and type of stress, it is necessary to determine the region where stress has no effect, the region where the effect is reversible, the region where stress has an irreversible effect and, possibly, the region of fracture.

The main mechanical stresses applied to the cable, particularly at the time of installation, are tension, crushing, shock, vibrational flexing, bending, torsion and vibration.

6.4. Thermal behaviour

The thermal behaviour of optical cables depends on the cabling process and the choice of materials. Thermal cycling, applied to these cables, enables inauspicious stresses and the existence of relaxation and ageing effects to be revealed.

7. THE PARAMETERS OF OPTICAL CABLES

Cable parameters divide into those of the cable itself and those of the fibre.

7.1. Cable parameters

— structure: grooved, compact;
— external diameter;
— standard length;
— weight in kg/km;
— operating temperature range;
— tensile strength;
— resistance to crushing;
— minimum radius of curvature;
— concentricity;
— imperviousness;
— fire resistance;

— use: video or digital
           internal, external, underground, free air...;
— number of fibres.

7.2. Fibre parameters

— type: graded index, step index, monomode...;
— core diameter;
— cladding diameter;
— concentricity error;
— non-circularity of the core;
— numerical aperture;
— attenuation for different wavelengths;
— bandwidth for different wavelengths;
— diameter of primary coating;
— nature of primary coating;
— tensile strength;
— standard length;
— diameter of secondary sheath;
— nature of secondary sheath.

A cross section of the cable is generally provided which enables the location of the various constituents to be determined.

8. INSTALLATION OF OPTICAL FIBRE CABLES

There are two cases to be considered; installation of buried cables and installation in free air.

8.1. Buried cables

Subterranean cables must be protected against damage which could be caused by subsidence of the earth, contact with hard objects or the shock of mechanical tools.
Two processes are used:

(a) complete burial of the cable observing conventional installation practices (laying depth, provision of alarm devices and so on) but without the need for particular protection;
(b) installation in a sheath in which case the internal diameter of the sheath

must be sufficiently greater than the diameter of the cable:

— diameter of the cable × 1.5 for a single cable and
— diameter of the cable × 2.5 for three cables.

8.2. Cables in free air

Only cables with a suitable coating can be installed in free air or internally; other cables must be protected by appropriate conduits. Each type of installation has its own characteristics; in particular the following arise:

— installation along walls;
— installation in cable runs or panels;
— installation in ducts.

In the case of aerial installation on a supporting cable, the optical cable is, most often, coiled round the support.

For vertical installation, it is sometimes necessary to provide several intermediate loops to avoid the possibility of the optical fibres slipping in the cable.

8.3. Mounting

The cable can be damaged during installation for various reasons such as:

— excessive tension;
— torsion;
— impact;
— cutting, scraping etc.

These should be considered before routeing the cable in order to minimize the risk of damage.

Finally, one danger is to fail to differentiate between optical and electrical cables. This can cause confusion such as inadvertant cutting of an optical fibre cable. It is thus preferable to distinguish the optical cable from other existing cable types.

7

COUPLING OF OPTICAL FIBRES

1. INTRODUCTION

In general, optical fibres as manufactured have lengths of 1.5–2 km, which is very often insufficient for installation on site. It is, therefore, necessary to join two optical fibres but losses at the connection must be minimised in order to preserve the dynamic range of the link.

Two classes of coupling can be distinguished — those described as passive, which cannot be disassembled (such as bonding and splicing), and those described as active, which can be disassembled (such as connectors and demountable splices).

Coupling losses can be separated into those associated with the nature of the fibre (intrinsic parameters) and those associated with the connection itself (extrinsic parameters).

2. INTRINSIC PARAMETERS WHICH CAUSE LOSSES

Manufacturing tolerances incur variations in fibre parameters which cause connection losses. In this context, the parameters concerned are:

— the core diameter;
— the numerical aperture;
— the refractive index profile.

By convention, the fibre by which the radiation arrives is described as the 'transmitter' and that which receives the radiation as the 'receiver'.

2.1. Losses associated with the core diameters

Consider two fibres, the transmitter, of core diameter a_t, and the receiver, of core diameter a_r; the losses at the interface are related to the common surfaces. If $a_r > a_t$, there is clearly no loss. In the other case:

$$L_d = 10 \cdot \log \frac{\pi a_r^2 / 4}{\pi a_t^2 / 4} \tag{1}$$

$$L_d = 20 \cdot \log \frac{a_r}{a_t} \qquad a_t > a_r \tag{2}$$

Fabrication tolerances give diameter accuracies of less than 5%, which cause losses of less than 0.42 dB.

2.2. Losses associated with numerical aperture

Let NA_t be the numerical aperture of the transmitter and NA_r that of the receiver; here again, the only case which causes a coupling loss is that where $NA_t > NA_r$. The variation in numerical aperture is related to the cross section of the core:

$$L_{NA} = 20 \cdot \log \frac{NA_r}{NA_t} \qquad NA_t > NA_r \tag{3}$$

The trend and magnitude of this loss are similar to those of losses due to variations in diameter.

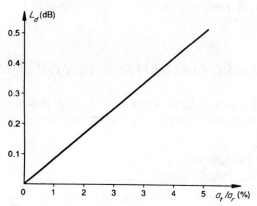

Figure 7.1 Losses associated with diameter discrepancies

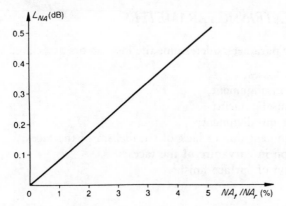

Figure 7.2 Losses associated with difference in numerical aperture

2.3. Losses associated with refractive index variation

Losses due to differences in refractive index can be ignored when the refractive index coefficient is greater on the receiving side than the transmitting side, that is:

$$L_\alpha = 10 \cdot \log \frac{1 + 2/\alpha_r}{1 + 2/\alpha_t} \qquad \alpha_t > \alpha_r \qquad (4)$$

For step index fibres, the refractive index parameter is of the order of 100 and these losses can be neglected.

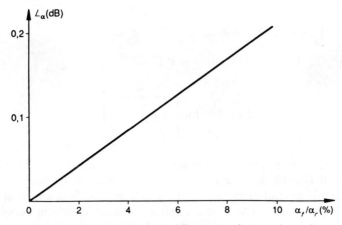

Figure 7.3 Losses associated with differences in refractive index profile

3. EXTRINSIC PARAMETERS

The extrinsic parameters determine the loss factors associated with coupling:
— Fresnel losses;
— radial misalignment;
— axial misalignment;
— angular misalignment;
— misalignment due to lack of parallelism of the faces;
— variation in curvature of the faces;
— variation of surface finish.

3.1. Fresnel losses

Transition from a medium of refractive index n_1 to a medium of refractive index n_2 always causes a loss of flux due to reflection of part of the flux at the interface.

From Fresnel's law

$$P_t = 1 - \left[\frac{n_1 - n_0}{n_1 + n_0}\right]^2 \tag{5}$$

where P_t is the flux transmitted. Hence

$$P_t = \frac{4}{2 + \dfrac{n_0}{n_1} + \dfrac{n_1}{n_0}} \tag{6}$$

the Fresnel losses can be written

$$L_F = 10 \cdot \log(P_t) \tag{7}$$

$$L_F = 10 \cdot \log\left[\frac{4}{2 + \dfrac{n_0}{n_1} + \dfrac{n_1}{n_0}}\right] \tag{8}$$

For a typical value of $n_1 = 1.5$ for silica and $n_0 = 1$ for air, the Fresnel losses amount to 0.177 dB. It is necessary to take account of transition from the transmitter to air and from air to the receiver, that is a coupling loss of 0.35 dB. To minimize these losses, a refractive index matching liquid is used with $n_0 \approx 1.5$; the losses then become negligible.

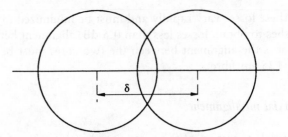

Figure 7.4 Radial offset

3.2. Radial misalignment

When the cores of the transmitter and receiver are separated by δ with respect to identical diameters D, losses arise due to misalignment of the axis. These are related to the variation in cross section.

It can be shown that the ratio of common cross sections of the two fibres is

$$R = \frac{2}{\pi} \left[\cos^{-1}\left(\frac{\delta}{D}\right) - \frac{\delta}{D} \left(1 - \left(\frac{\delta}{D}\right)^2\right)^{1/2} \right] \quad (9)$$

The radial losses can be written

$$L_{ra} = 10 \cdot \log(R)$$

hence

$$L_{ra} = 10 \cdot \log\left[\frac{2}{\pi} \left[\cos^{-1}\left(\frac{\delta}{D}\right) - \frac{\delta}{D} \left(1 - \left(\frac{\delta}{D}\right)^2\right)^{1/2} \right]\right] \quad (10)$$

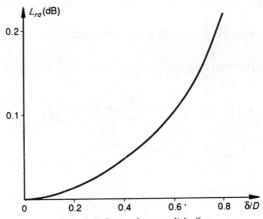

Figure 7.5 Losses due to radial offset

Notice that these losses vary rapidly and must be minimized in the connectors. If one wishes to obtain losses less than 0.5 dB, alignment better than 0.1 is required; that is the alignment between the two cores must be better than 5 μm for a 50/125 μm fibre.

3.3. Axial misalignment

Losses due to axial separation of the cores can also arise. As a first approach to analysis, it will be assumed that the illumination in the receiving plane is uniform and that the proportion of incident power injected into the fibre is equal to the proportion of the cross section illuminated.

Let D be the core diameter of the fibres, α_L the aperture angle and d the distance between the two fibres.

$$L_{ax} = 10 \cdot \log \left[\frac{\dfrac{\pi D^2}{4}}{\pi \left(\dfrac{D}{2} + d \cdot \tan(\alpha_L) \right)^2} \right] \tag{11}$$

$$L_{ax} = 20 \cdot \log \left[1 + 2 \, \frac{d}{D} \tan(\alpha_L) \right] \tag{12}$$

Here also, losses are very significant since for a ratio of 0.5 there is a loss of 1.8 dB.

3.4. Angular misalignment

The last alignment error is angular deviation between the two ends of the fibre.

In this case, it can be shown that the associated loss is

$$L_{ang} = 10 \cdot \log \left[\cos(\phi) \left(\frac{1}{2} - \frac{1}{\pi} \, p(1 - p^2)^{1/2} - \frac{1}{\pi} \sin^{-1}(p) \right) \right.$$
$$\left. + q \left(\frac{1}{\pi} \, r(1 - r^2)^{1/2} + \frac{1}{\pi} \sin^{-1}(r) + \frac{1}{2} \right) \right] \tag{13}$$

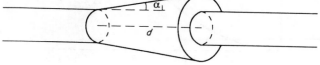

Figure 7.6   Axial shift

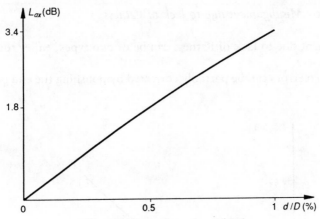

Figure 7.7 Losses due to an axial shift

where:

$$p = \frac{\cos(\alpha_L)(1 - \cos(\phi))}{\sin(\alpha_L) \cdot \sin(\phi)}$$

$$q = \frac{\cos^3(\alpha_L)}{(\cos^2(\alpha_L) - \sin^2(\phi))^{3/2}}$$

$$r = \frac{\cos^2(\alpha_L)(1 - \cos(\phi)) - \sin^2(\phi)}{\sin(\alpha_L) \cdot \cos(\alpha_L) \cdot \sin(\phi)}$$

$$\sin(\alpha_L) = NA$$

A mounting error of a few degrees is quite common, but causes a loss of less than 0.5 dB, which is less significant than the other alignment errors.

3.5. Misalignment due to non-parallelism of the faces

The ends of the fibre are not always orthogonal.

From an analytic point of view, this can be considered as an angular misalignment.

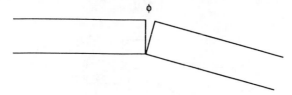

Figure 7.8 Angular offset

3.6. Misalignment due to lack of flatness

Misalignment due to lack of flatness can be of two types, either roughness or convexity.

These two errors can be partially corrected by polishing the end of the fibre.

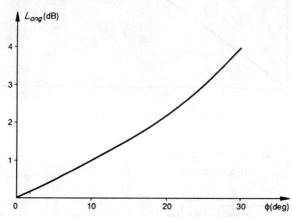

Figure 7.9   Losses due to an angular offset

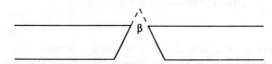

Figure 7.10   Offset due to non-parallelism

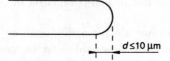

Figure 7.11   Convexity imperfection

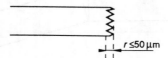

Figure 7.12   Roughness imperfection

4. PREPARATION OF THE FIBRE ENDS

It has been seen that three types of loss are associated with the surface state of the fibre end:

— the orthogonality of the face with respect to the axis of the fibre;
— convexity;
— roughness.

The quality criteria are a surface which is as flat as possible, orthogonal to the axis of the fibre and of optical polish. Two techniques enable the ideal state to be approached — cleaving and polishing.

4.1. Cleaving

The advantage of this technique is its speed, its low cost and the absence of residue on the sectioned face after cleaving; the drawback is the difficulty of obtaining a 'good' break.

After cleaving a fibre, one can generally distinguish three areas on the face obtained — the mirror, the intermediate region and the comb.

The difficulty of the method is to obtain a cleaved face which has only the aspect of a mirror. The first stage is to initiate a break with a diamond point which must not in any event reach the core.

The second step is to bend the fibre to open the fracture by exerting a slight force on each end.

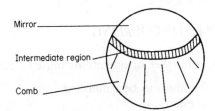

Figure 7.13   A cleaved face

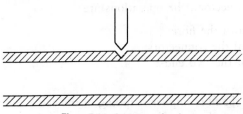

Figure 7.14   Initiating a break

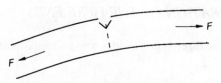

Figure 7.15 Producing a cleave

During this operation, the initial fracture propagates and provides a clean cleavage of the fibre. If the operation is not perfectly successful, the main faults are splintering of the end, lack of orthogonality and convexity.

4.2. Polishing

This operation is performed when the fibre is mounted in the connector. It is then secured in a calibrated polishing chuck which rotates with a random movement on abrasive discs. The chuck also serves as a gauge so that the length of the fibre in the connector is optimized for the coupling.

The polishing operation generally uses three different grits; the final finish is provided by the finest, which has a diameter of 0.3 μm and enables a mirror finish to be obtained.

The advantage of this method is that good orthogonality of the face can be obtained and impairments due to convexity and roughness are small; the major disadvantage is the number of stages and the duration of each. This technique is rarely used except for connectors.

5. PRINCIPLES OF COUPLING

During a connection, a number of operations must be performed:

— preparation of the fibres to be connected;
— performing the connection;
— protecting the connection.

In the case of connectors, the operations are:

— preparation of the fibre;
— mounting of the connectors;
— polishing of the two ends;
— coupling.

Preparation of the fibre implies stripping of the cable and the coating from the fibre in order to obtain a fibre consisting only of the core and cladding.

Once the connection is performed, it is necessary to provide adequate protection, as part of the installation, in order not to leave any weak points in the link.

6. CONNECTION BY WELDING (FUSION)

Welding is the connection technique which provides the least loss, since this is typically less than 0.1 dB. Firstly, the two ends of the cleaved optical fibres are aligned under a microscope; then welding is performed by means of an electric arc.

The disadvantages of the method are the amount of hardware needed which impedes its use in some sites (such as cable runs), the difficulty of obtaining a 'good' weld and the non-demountable nature of the connection.

7. SPLICES

Demountable or fixed, splicing techniques involve bringing two cleaved fibres into contact in a bath of refractive index matching liquid.

Single-fibre splices generally consist of a Vee form fibre guide filled with refractive index matching liquid into which the fibres are slid until they are in contact. The fibres have previously been stripped and cleaved.

According to the particular case, a plate or other device is installed to protect the splice; the cable is then reformed in such a way as not to leave a weak point in the link.

Some gels which are used are polymerizable; in this case the demountable property of the link is lost.

With techniques of this kind, losses are less than 0.3 dB; this is the technique most commonly used for medium-distance links (around 10 km).

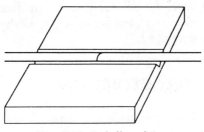

Figure 7.16   Single-fibre splicing

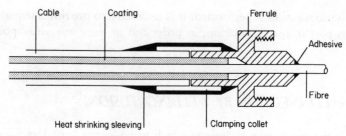

Figure 7.17   Diagram of a connector

8. CONNECTORS

Although sometimes used in the link itself, connectors are above all used at the transmitter–fibre and fibre–receiver interfaces.

Two independent parts can generally be identified. The first is the core of the connector in which the fibre to be connected is precisely positioned; the second is the cover which includes a cable clamp and a system for locking the connector.

The cable is stripped in two regions, one as far as the cladding (20–30 mm) and the other including the coating (10 mm). The fibre is coated with adhesive and introduced into the connector which has a guiding cone where the stripped fibre enters but the coated fibre is blocked. The cable reaches a stop and is clamped to the connector before being supported by a heat-shrinkable sheath. After ensuring that the adhesive has passed through the connector to the fibre exit, the adhesive is heated to polymerize it. The fibre is fractured where the adhesive has emerged and then polished until a mirror surface is obtained.

The loss is generally less than 1 dB and can be reduced by the use of refractive-index-matching liquid. The main disadvantages are the time to mount the connector (one hour), the magnitude of the losses and the non-demountability of the connector. The advantages are the robustness of the link, the possibility of repolishing and the demountability of the link.

Connectors of all types for all applications are found (insulating, rust-proofed, impervious and so on), but for present links the standard Amphenol 906 has been highly developed.

9. CONNECTION SPECIFICATION

Specification of a connection is made in accordance with the fibre, the cable and the type of connection.

Fibre   — monomode or multimode;
               — core/cladding.
Cable   — external diameter.
Connection   — fixed or demountable;
               — criticality of losses;
               — facility for installation on site;
               — environment (insulating, marine and so on).

In practice, there is a tendency to use standard easily mounted connectors, but, in the case of multi-channel cables it is necessary to use special techniques.

10. PASSIVE COUPLERS

10.1. Introduction

In configuring a serial network, it is useful to be able to transmit the information in a loop while enabling each element to transnit or receive (Fig. 7.18). To apply this technology to a fibre optic network, passive serial couplers are used.

Let P_0 be the power injected into the coupler; a power P_t is transmitted with coupling to the station receiver of a power P_c and rejection to the transmitter of a power P_r.

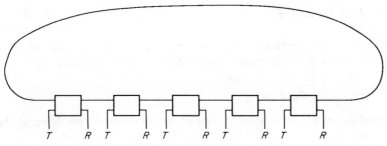

Figure 7.18   Serial distribution

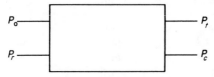

Figure 7.19   Serial coupler

On passing through each coupler, there is therefore a loss which requires the use of automatic gain control at each receiver.

Transmission loss

$$A_t = 10 \cdot \log \frac{P_t}{P_0}$$

Coupling loss

$$A_c = 10 \cdot \log \frac{P_c}{P_0}$$

Rejection loss

$$A_r = 10 \cdot \log \frac{P_r}{P_0}$$

Total insertion loss

$$A_i = 10 \cdot \log \frac{P_t + P_r + P_c}{P_0}$$

Table 7.1 shows the coupling performance of an example of serial coupling; notice that in this case the inputs have rejections of 40 dB and the transmitted part is equal to the coupled part. Other ratios can be obtained to suit each network; it is also necessary to include the coupling losses of the serial coupler to the network, and this depends on the type of coupling chosen.

10.2 Realization

There are several methods of realizing passive couplers; two which are currently used will be described: fusion-drawing and polishing.

10.2.1. Fusion-drawing

This method enables symmetrical couplers to be realized with N input chan-

Table 7.1 Interaction of inputs

dB	Input	Trans.	Coup.	Rejected
Input	40	3.5	3.5	40
Trans.	3.5	40	40	3.5
Coup.	3.5	40	40	3.5
Rejected	40	3.5	3.5	40

nels and N output channels where N can vary from 2 to 100 for different coupling coefficients.

The N fibres are twisted and the common region is heated to the softening temperature of glass; this enables the fibres to be drawn and then fused into a coupler.

Here also, the coupling coefficient is greater for the higher modes than for the other modes and account must be taken of this phenomenon in cascade configurations.

10.2.2. Polishing

This technique consists of bending a fibre and polishing it laterally to expose the core to a particular depth (Fig. 7.20).

The two fibres prepared in this way are then mounted facing each other with a refractive-index-matching liquid. The coupler is formed into a single unit by bonding into a thin layer (Fig. 7.21).

This type of technique is applicable only to four channel couplers but has the advantage of being applicable to all types of fibre. The long-term stability depends on the bonding system used, typical losses are less than a dB.

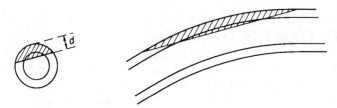

Figure 7.20 Polishing a fibre (the hatched part is removed)

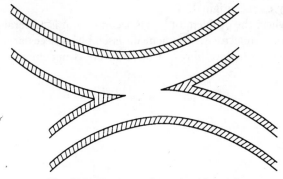

Figure 7.21 Passive coupler produced by polishing

11. ACTIVE COUPLERS

These are electronic systems which receive the signals from several channels and redistribute them in other channels. This point will be developed in the context of links using repeaters.

12. MISCELLANEOUS COUPLING SYSTEMS

Other functions can be required in optical fibre links and some will be indicated here.

One of the more common systems is a wavelength multiplexer which enables several wavelengths to be transmitted on the same fibre.

'Choppers', which are switches for fibre optic links, can be controlled by an external electrical signal and permit control of the opening and closing of links.

There is a family of fixed or controllable diverters which enable the optical path of a network to be modified.

There are also variable attenuators; these are suitable for control and dynamic examination of networks.

13. CHOICE OF COUPLING

During the design of a fibre optic link, it is necessary to determine the type of fibre coupling to be used. The choice is generally a compromise between the acceptable loss rate, the facility for assembling and disassembling the coupling, and its reliability.

In a case where losses are critical, the choice will be a weld (loss ≈ 0.1 dB), in spite of the realization difficulties and the need to reform a cable with the same reliability as the original.

In the case where the problem of loss is secondary, it is preferable to choose a coupling using connectors if frequent disassembly is necessary, or splicing, with the provision of spare fibres, if disassembly is uncommon.

8

THE LIGHT-EMITTING DIODE (LED)

1. THE PRINCIPLE OF LIGHT EMISSION IN A SEMICONDUCTOR

1.1. Introduction

In a semiconductor, electrons can make a transition between the valence band and the conduction band either if the forbidden band (the gap) is not too wide or in the presence of impurities which create intermediates levels in this gap. The energy required by the electrons to pass from one level to another is at least equal to the energy of the forbidden band:

$$Ec - Ev = Eg \tag{1}$$

These transitions correspond to a recombination of levels and this tends to fill up the holes in the valence band with electrons from the conduction band.

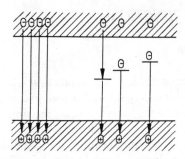

Figure 8.1 Transitions in a semiconductor

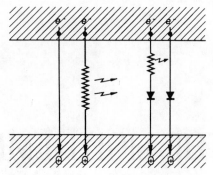

Figure 8.2 Radiative and non-radiative transitions

One method of causing these transitions is to create holes artificially in the valence band by doping. A transition of electrons from the conduction band to the valence band is then caused by an external application of energy (external biasing); this is the case with semiconductors.

1.2. Radiative and non-radiative transitions

This type of transition can be either non-radiative, in which case there is no emission of light, or radiative. In the latter case radiation is emitted and this is the case with light-emitting diodes (LEDs). Radiative transitions can be natural or stimulated; in the case of the diode laser, the transition is stimulated.

The frequency of the emitted radiation is given by

$$hf = Ec - Ev \qquad (2)$$

where h is Planck's constant ($h = 6.626 \times 10^{-34}$ J/s).

2.   THE p–n JUNCTION

2.1. Introduction

The basis of a light-emitting diode is a p–n junction which consists of a semiconductor having an n-doped region and a p-doped region in contact. The transition from one region to the other is linear; the simplest case is the homojunction.

At the junction, an electrical equilibrium is created due to the simultaneous presence of charge carriers and electrons in equal numbers. The difference in

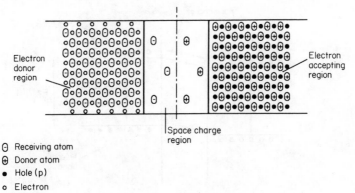

Figure 8.3   A p–n junction

energy level between the valence band and the Fermi band results from doping; E_i is the energy of the impurities responsible for the radiative transitions.

When the junction is formed, the Fermi levels are aligned (Fig. 8.4).

In equilibrium, the potential well is too deep to allow electrons from the n region to make a transition to the p region.

2.2. Operation

Biasing the junction in the forward direction by an external potential V reduces the potential barrier and this causes a reduction in the space-charge region and permits electrons to make a transition from the n region to the p region (Fig. 8.5).

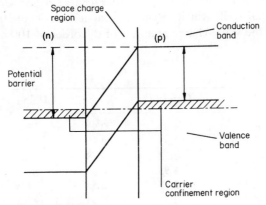

Figure 8.4   Alignment of the Fermi levels

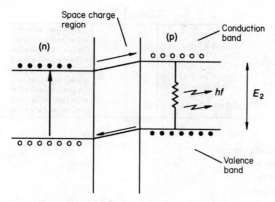

Figure 8.5 Modification of the levels by a potential

Some of the electrons arriving in the conduction band from the p region make a transition to the valence band of the n region by giving rise to a radiative transition; this is the principle of the light-emitting diode.

2.3. Materials

The radiation depends on the width of the forbidden band which in turn depends on the type of material used (Table 8.1).

2.4. Single and double heterojunctions

This structure involves surrounding the space-charge region with a confining region (a potential well) which retains the carriers in defined regions.

The width of the space-charge region is of the order of 100 nm and that of

Table 8.1

Material	Gap (eV)	λ (μm)	Colour
GaAs	1.3	0.95	I.R.
GaAs(Zn)	1.37	0.90	I.R.
GaAlAs	1.82	0.68	Red
GaAsP	1.93	0.64	Red
GaAsP(n)	2.13	0.58	Yellow
GaP	2.3	0.54	Green

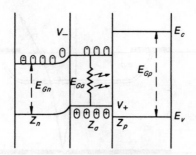

Figure 8.6 The basis of a heterojunction

the confining potential well is 50 nm. The heterojunction is described as double when the space-charge region is confined between two potential wells; it is single when the confinement is unilateral. The advantage of this type of structure is that it ensures that transitions are confined to a very small region and this increases the efficiency within a given solid angle.

3. STRUCTURES

Light-emitting diodes consisting of a p–n junction are produced from substrates which can be either absorbent or transparent.

3.1. Absorbant substrates

Rays which propagate in the substrate are not reflected but absorbed; the efficiency of this type of system is rather low ($\rho \approx 1.3\%$).

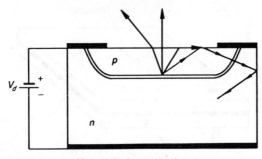

Figure 8.7 A p–n junction

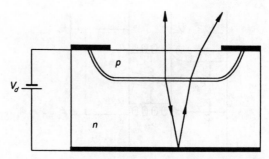

Figure 8.8   A p–n junction with transparent substrate

3.2. Transparent substrates

The structure is similar to the previous case but the substrate does not absorb the rays; these are reflected from the bottom of the substrate which is covered with a reflecting contact and this improves the efficiency.

3.3. The emission field

A light-emitting diode can, in general, emit in three different directions. An emission direction perpendicular to the junction is usually chosen since the absorption coefficient of the material is sufficiently low to permit emission of a large number of photons.

If the active region of the diode is surrounded by an optical guiding region, the created photons are confined in a single region and in a very low field; this increases the photon density per unit solid angle. This principle is used for high-efficiency diodes and coupling with an optical fibre.

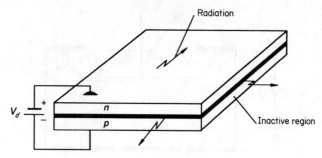

Figure 8.9   The emission field

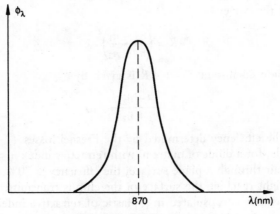

Figure 8.10   Emission spectrum

3.4. The emission spectrum

The wavelength of the emitted radiation depends on the width of the forbidden band of the material,

$$f = \frac{Ec - Ev}{h} = \frac{c}{\lambda} \tag{3}$$

where f is the frequency emitted.

This emission is not strictly monochromatic for two main reasons:

— The presense of impurities in the material replaces the discrete levels of the material with energy bands.
— All the emitted wavelengths are not transmitted in the same way in the material or with the same amplitude.

Also, as the temperature increases, the frequency of emission moves towards the infra-red.

4. THE EFFICIENCY OF A DIODE

4.1. The definition of optical efficiency

4.1.1. The Fresnel loss

Passage of light from one medium to another of different refractive index involves losses at the interface, called Fresnel losses. Let R be the reflection coefficient of the wave:

$$R = \left(\frac{n_2 - n_1}{n_1 + n_2}\right)^2 \tag{4}$$

the transmission coefficient $T = 1 - R$ is given by

$$T = \frac{4n_1 n_2}{(n_1 + n_2)^2} = \rho_F \tag{5}$$

where ρ_F is the efficiency determined by the Fresnel losses.

For example, for a diode of material with refractive index $n_1 = 3.4$ emitting directly into air through a plane surface, the efficiency is 70%. That is, only 70% of the light reaching the surface of the chip is transmitted into the air. The chip is usually encapsulated in a plastic of refractive index 1.5 and the overall efficiency can be written

$$\rho_T = T_1 \cdot T_2 \tag{6}$$

$$\rho_T = \frac{4n_1 n_x}{(n_1 + n_x)^2} \cdot \frac{4n_x n_2}{(n_x + n_2)^2} \tag{7}$$

where n_x is the refractive index of the intermediate material ($n_x \approx 1.5$); the efficiency becomes 82%, an improvement of 10%, which can be further improved.

4.1.2. Efficiency of the critical angle

The passage of light from one medium to another is governed by the law of refraction; this defines a limiting angle which, when exceeded, causes reflection into the transmitting medium. Snell's law gives

$$n_1 \cdot \sin(\theta_1) = n_2 \cdot \sin(\theta_2) \tag{8}$$

the critical angle is defined by $\theta_2 = 90°$.

$$\sin(\theta_c) = n_2/n_1 \tag{9}$$

$$\theta_c = \arcsin(n_2/n_1) \tag{10}$$

with the condition $n_2 < n_1$.

The following ratio is called the efficiency of the critical angle:

$$\rho_{cr} = \left(\frac{n_2}{n_1}\right)^2 \tag{11}$$

In air, with $n_1 = 3.4$ and $n_2 = 1$, one obtains 8.6%. In the presence of an intermediate material of different refractive index ($n_x = 1.5$), one obtains an increase of efficiency to 19%.

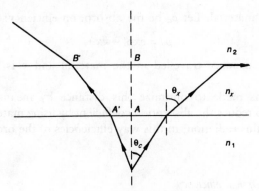

Figure 8.11   Spreading of the beam

Another property of the intermediate medium is spreading of the output
beam.

In the presence of an intermediate dielectric layer, the critical angle does not
vary but the beam is enlarged. In fact the distance AA' becomes BB'; this
increase is proportional to the thickness of the dielectric.

In order to increase the output beam to a maximum, the flat intermediate
medium is replaced by a hemisphere.

In the case of AsGa, the critical angle changes from $17°$ to $26°$ and the
efficiency from 0.08 to 0.19 or approximately double.

4.1.3. Absorption efficiency

During its passage through the material, the radiating flux from the junction
is absorbed in accordance with an exponential law in terms of the distance

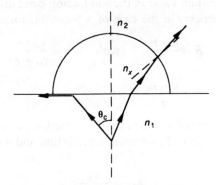

Figure 8.12   Intermediate medium in a hemisphere

travelled in the material. Let ρ_a be the absorption efficiency:

$$\rho_a = \exp(-kx) \tag{12}$$

where x is the distance travelled in the material and k is the absorption constant.

An attempt is made to optimize this distance by means of the diode geometry and to reduce the absorption coefficient by using material with a low absorption for this radiation; in this way efficiencies of the order of 75% are achieved.

4.1.4. Optical efficiency

The product of the various efficiencies associated with optical losses is called the optical efficiency.

$$\rho_{0\rho} = \rho_a \cdot \rho_E \cdot \rho_{cr} \tag{13}$$

absorption Fresnel critical angle

The usual order of magnitude of the diode efficiency is 15%.

4.2. Quantum efficiency

4.2.1. Quantum efficiency of luminescence

The quantum efficiency of luminescence $\rho_q L$ is defined by the ratio of the number of radiative recombinations to the total number of recombinations:

$$\rho_{qL} = \frac{R_r}{R_T} \tag{14}$$

The rate of recombination R is proportional to the number of carriers exceeding the equilibrium value in the conduction band divided by the lifetime of this excess of carriers. In the case of a p region, for example:

$$R = \frac{n - n_0}{\tau} \quad \text{with} \quad \tau = \frac{\tau_{nr} \cdot \tau_r}{\tau_{nr} + \tau_t} \tag{15}$$

$$R = \frac{n - n_0}{\tau_r} \tag{16}$$

where τ_{nr} is the lifetime of a non-radiative transition and τ_r that of a radiative transition, n_0 the number of carriers in equilibrium and n the total number of carriers.

$$\rho_{qL} = \frac{\tau_{nr}}{\tau_{nr} + \tau_r} \tag{17}$$

For this efficiency to tend to 1, it is necessary for the lifetime of the radiative transition to be as short as possible. This efficiency, usually 40%, decreases with temperature.

4.2.2. External quantum efficiency

The quantum efficiency defines the number of photons emitted by injected electrons:

$$\rho_{qe} = \frac{\text{photons}}{\text{electrons}} \qquad (18)$$

Let I be the current through the diode; then

$$I = n_e \cdot q \qquad (19)$$

where q is the charge on the electron and n_e the number of electrons. The luminous power emitted by the diode is expressed in watts such that

$$\phi = n_\rho \cdot E_g \qquad (20)$$

where $E_g = hc/\lambda$ and n_ρ is the number of photons emitted. The external quantum efficiency can then be written

$$\rho_{qe} = \frac{\phi}{I} \cdot \frac{q\lambda}{hc} \qquad (21)$$

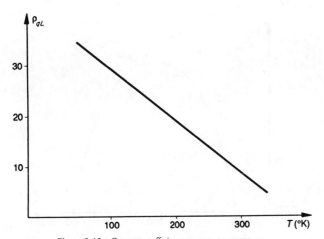

Figure 8.13 Quantum efficiency versus temperature

which is usually written

$$\rho_{qe} = \frac{\phi}{I} \cdot \frac{\lambda \ (\text{nm})}{1240} \tag{22}$$

where ϕ is expressed in watts and i in amperes.

4.2.3. Luminous efficiency

In optics, a distinction is made between what is seen by the human eye, which is measured in lumens, and the luminous power, in the physical sense of the term, which is measured in watts.

A curve of relative luminosity has been defined which enables the two systems of units to be related in such a way that at the maximum sensitivity of the eye 1 watt = 680 lumens.

The total flux (in watts) of a diode, as a function of wavelength, can be written

$$\Phi_T = \int_0^\infty \frac{\mathrm{d}\phi}{\mathrm{d}\lambda} \cdot \mathrm{d}\lambda \tag{23}$$

By taking account of the sensitivity of the eye, the expression for the flux (in lumens) with respect to the eye can be written

$$\phi_L = 680 \int_0^\infty \frac{\mathrm{d}\phi}{\mathrm{d}\lambda} \cdot V_\lambda \cdot \mathrm{d}\lambda \tag{24}$$

where V_λ defines the sensitivity curve of the eye. The luminous efficiency of

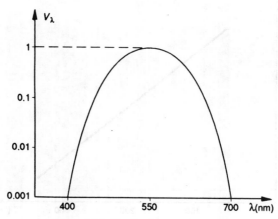

Figure 8.14   Watt/lumen conversion curve

Table 8.2

	λ (nm)	$\rho v(\text{lm}/\text{W})$
Standard red	655	60
High efficiency red	655	135
Yellow	585	540
Green	565	640

a light-emitting diode is then defined with respect to the eye by the ratio

$$\rho v = \frac{\Phi_L(\text{lm})}{\Phi_T(W)} \tag{25}$$

Typical values of efficiencies of diodes are as shown in Table 8.2.

5. FLUX AND LUMINOUS INTENSITY

5.1. Radiant intensity

The radiant intensity of a diode is defined with respect to a solid angle as follows:

$$I_e(\text{w}/\text{Sr}) = \frac{I_V(\text{cd})}{\rho_V(\text{lm}/\text{W})} \tag{26}$$

since 1 cd = 1 lm/Sr, I_e is expressed in watts per steradian. In the case of a high efficiency red diode which emits 12 mcd for example, one obtains

$$I_e = \frac{12 \cdot 10^{-3}}{135} = 89 \ \mu\text{W} \tag{27}$$

5.2. Calculated luminance

The radiation diagram of a diode with respect to angle can be plotted in polar or cartesian coordinates. The luminous intensity on the diode axis I_0 is measured in millicandelas (mcd) and can be as much as a candela. The total flux emitted by a diode over all space is

$$\phi = \int_0^\pi I(\theta) \cdot 2\pi \cdot \sin(\theta) \ d\theta \tag{28}$$

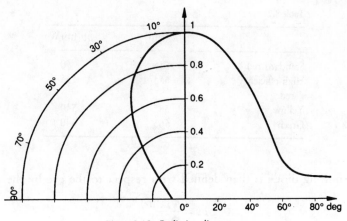

Figure 8.15   Radiation diagram

Let I_0 be the intensity emitted on the diode axis; the relative intensity I_r such that $I_r(0) = 1$ enables one to write

$$\phi = I_0 \int_0^\pi I_r(\theta) \cdot 2\pi \cdot \sin(\theta) \, d\theta \qquad (29)$$

The luminous flux depends on the type of source, for example:
for a point:

$$I_r(\theta) = 1, \qquad \phi = 4\pi I_0 \qquad (30)$$

for a Lambertain source:

$$I_r(\theta) = \cos(\theta), \qquad \phi = \pi I_0 \qquad (31)$$

For a diode, the following is used as a function representing the source:

$$I_r(\theta) = \cos^n(\theta) \qquad (32)$$

with

$$\theta_{1/2} = \frac{1}{\cos\left(\dfrac{1}{2^{-n}}\right)} \qquad (33)$$

where n is a parameter which depends on the diode; the flux is then written as

$$\phi = \frac{2\pi}{n+1} I_0 \qquad (34)$$

5.3. Magnification and luminous intensity

5.3.1. Principle

The luminous intensity of a diode depends on the magnification provided by the encapsulating lens. If the surface of the sphere is defined as a principal plane, a focal point can be defined as a point from which parallel rays are emitted at the output of the lens; the expression for this focal distance is

$$f = \frac{r}{1 - \dfrac{n_2}{n_1}} \tag{35}$$

where r is the radius of the diode.

The focal length (f) is determined by the radius of curvature corrected by a factor depending on the refractive indices of the two media. The magnification m is calculated from the distance between the object (in this case the emitting semiconductor) and the surface of the sphere.

$$m = \frac{1}{1 - \dfrac{x}{f}} \tag{36}$$

In the case where the distance $x = 0$, there is no magnification, the light source is Lambertian and hence the luminous flux is πI_0, where I_0 is the luminous intensity emitted on the axis.

Provided that x is assumed to be between 0 and r, the magnification and luminous intensity are related by

$$I = m^2 \cdot I(\theta) \tag{37}$$

where $I(\theta)$ is the characteristic of the Lambertian source.

In the case where x is greater than r, the source is no longer Lambertian, the luminous intensity along the axis increases and the total luminous flux decreases.

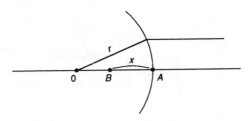

Figure 8.16 Definition of the focal length of a diode

5.3.2. Diodes with and without a diffuser

In order to increase the aperture angle of the luminous flux, a diffuser which disperses the light is introduced (Figs. 8.17, 8.18).

Diodes used for optical fibres are equipped with miniature lenses which ensure convergence of the beam compatible with the numerical aperture of the fibre (Fig. 8.19).

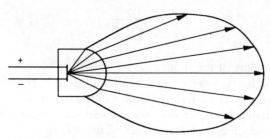

Figure 8.17 A diode in a plastic package without a diffuser

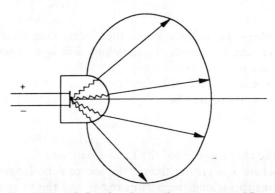

Figure 8.18 A diode with a plastic package containing a diffuser

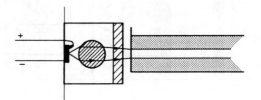

Figure 8.19 A diode for optical fibre use

6. OPTICAL PARAMETERS OF THE DIODE

Four main parameters are relevant in specifying a diode for optical fibre work:

— the emission spectrum;
— the radiation diagram;
— the numerical aperture;
— the diameter of the optical window.

6.1. The emission spectrum

This defines the variation of luminous flux from the diode as a function of wavelength and enables the wavelength of peak emission to be determined (see Fig. 8.10); this is the wavelength for which luminous emission is a maximum. This peak should be as narrow as possible to avoid monochromatic dispersion.

6.2. The emission diagram

This characterizes the distribution of luminous flux from the diode as a function of the angle of emission (see Fig. 8.15). For fibres, the maximum energy is emitted in the direction of the optical axis. This diagram also enables the half angle to be defined which corresponds to the angle for which the luminous intensity is halved with respect to the maximum intensity.

6.3. The numerical aperture

The numerical aperture corresponds to the half angle in which 90% of the luminous energy of the diode is emitted.

$$NA = \sin(\theta_{1/2}) \tag{38}$$

6.4. The diameter of the optical window

The output window of the diode is either a parallel faced plate or, more often, a microlens which ensures compatibility between the numerical aperture of the diode and that of the fibre. This parameter is important since it arises in the calculation of coupling losses and evaluation of the link.

7. ELECTRICAL AND THERMAL PROPERTIES OF THE DIODE

As for a conventional diode, the light-emitting diode is specified by its useful current and forward voltage. The passage of current through the diode causes a variation of temperature which depends on the mode of use (pulsed or continuous); this temperature variation causes changes in the optical and electrical characteristics of the diode.

7.1. Operating limits as a function of temperature

Various factors associated with its design limit the operating temperature of a diode. Firstly there is the package temperature above which the properties of the diffusing dome will not be maintained; secondly the junction temperature of the diode must not be exceeded.

The power dissipated in the diode depends on the current flowing through it, its forward voltage and its dynamic resistance; furthermore this power increases as the temperature increases. In general, the permitted power dissipation in a diode decreases rapidly with increasing ambient temperature.

The mean power dissipated by the diode can be expressed in the form

$$P_m = I_m(V_F + R_d(I_p - I_s)) \tag{39}$$

where

I_m = mean current
I_p = peak current

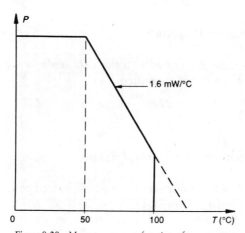

Figure 8.20   Mean power as a function of temperature

I_s = specified current
V_F = specified forward voltage
R_d = dynamic resistance of the diode.

For operation of the diode in continuous mode, it can be assumed that

$$I_p = I_s = I_m \qquad (40)$$

7.2. Operation in pulsed mode

In pulsed mode, the diode is not continuously illuminated; its peak and mean currents are therefore different. Consider the following example where the diode is required to be illuminated for 30% of a period and, in the present case, the repetition frequency is 1 kHz. Let τ be the duration of illumination:

$$\tau = 0.3 / 1000 = 300 \; \mu s$$

Observe from Fig. 8.21 that the ratio of maximum peak current to continuous current which is acceptable for $\tau = 300$ is 2.4:

$$\frac{I_p}{I_{Dc}} = 2.4 \qquad (41)$$

The operating current of the diode is 20 mA, the permitted peak current is

$$I_p = 2.4 \times 20 \; \text{mA} \qquad (42)$$

and the mean current is

$$I_m = \tau \cdot I_p \qquad (43)$$

the mean power dissipated is thus less than in continuous operation, as is to be expected.

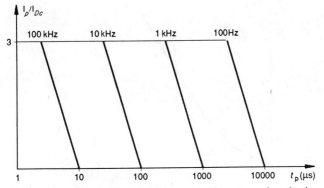

Figure 8.21 Maximum permissible peak pulse current with respect to the pulse duration

7.3. Mean luminous intensity

The axial luminous intensity, specified by the manufacturer, is generally obtained from photometric measurement when the diode is supplied with a specified continuous current: the value given can be typical or a guaranteed minimum. Other current levels provide other luminous intensities (ϕ):

$$\phi = \phi_{L0}\left(\frac{I}{I_F}\right)^n = \rho_c I \qquad (44)$$

where

ϕ_{L0} = reference luminous intensity
I_F = reference current
I = operating current
n = a characteristic of the diode which typically varies between 1.1 and 1.4
ρ_c = relative luminous efficiency for a given current.

Knowing the luminous intensity for any current, the mean luminous intensity in pulsed mode can be calculated:

$$\phi_m = \frac{I_p \cdot \tau \cdot r_p \cdot \phi}{I_c \cdot \rho_c} \qquad (45)$$

where r_p is the efficiency for the peak current.

Consider, for example, the case of a diode emitting 3 mcd at 16 mA; a peak current of 50 mA is required with a duty cycle of 30%, so

$$I_p = 50 \text{ mA} \qquad \phi_{L0} = 3 \text{ mcd}$$
$$\tau = 0.3 \qquad I_s = 16 \text{ mA}$$

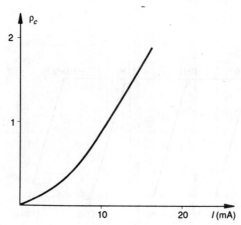

Figure 8.22   Relative luminous intensity as a function of current

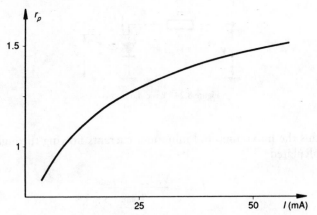

Figure 8.23 Relative luminous efficiency as a function of peak current

To define the two required efficiencies, the graphs of relative intensity versus current and relative efficiency versus peak current are used.

From these two graphs, it can be deduced that:

$$r_p = 1.5, \qquad \rho_c = 1.85$$

from which the mean luminous intensity is obtained:

$$I_m = 2.28 \text{ mcd} \tag{46}$$

8. ELECTRICAL PARAMETERS OF THE DIODE

The characteristic electrical parameters of the diode are the operating current (I_F) and the forward voltage (V_F) of the diode; these enable the electronic driving conditions to be defined.

8.1. Calculation of the load resistance

Because of its very low dynamic resistance, the diode cannot be connected in parallel with a resistance; this would cause too high a current to flow in the diode and an excessive power dissipation which would lead to its destruction.

The current flowing is thus limited by using a load resistance (R) for an operating voltage (V_c) as given by

$$I_F = \frac{V_c - V_F}{R} \tag{47}$$

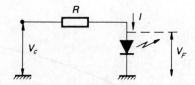

Figure 8.24   Basic diode circuit

From this the maximum and minimum currents flowing through the diode can be calculated:

$$I_{max} = \frac{V_{cmax} - V_{Fmin}}{R_{min}} \tag{48}$$

$$I_{min} = \frac{V_{cmin} - V_{Fmax}}{R_{max}} \tag{49}$$

In each case, the current must not cause damage to the diode.

8.2.   Example of a circuit with current regulation

Figure 8.25 shows a circuit diagram which does not provide control. The basic equations of this circuit are

$$R_1 = \frac{V_c - V_F}{I_b} \tag{50}$$

$$R_2 = \frac{V_2 - V_{be}}{I_c} \tag{51}$$

$$I = I_c + I_b \tag{52}$$

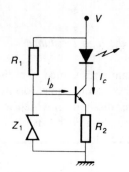

Figure 8.25   Circuit with current regulation

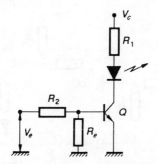

Figure 8.26   Digital control of a diode

With this circuit, the diode is continuously illuminated and cannot be turned off, on the other hand it is protected against variations in intensity by means of the zener diode which stabilizes the transistor base voltage.

If the control voltage V_i is zero, transistor Q is turned off and the current I_c through the diode is zero, it remains extinguished. In contrast, if the voltage $V_i > V_{be}$ and the base current is sufficient ($I_b > I_{bmin}$), the transistor turns on and the current in the diode is not zero; the diode illuminates.

In digital mode, the usual value of the control voltage is the supply voltage ($V_i = V_c$). The resistance R_e enables the base voltage to be fixed when $V_i = 0$; a typical value is 10 kΩ.

In this circuit, an npn transistor has been used which enables the diode to be illuminated when the control signal is in the high state. It is possible to make the diode operate with the control signal in the low state ($V_e = 0$); for this a circuit with a pnp transistor can be used.

These two types of circuit operate at low frequencies but must be more sophisticated above 200 kHz because of the effect of stray capacities which distort the signal.

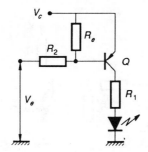

Figure 8.27   Digital control with a pnp transistor

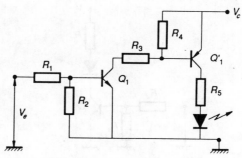

Figure 8.28   Cascade npn–pnp circuit

Circuits with current limits can be produced by using cascade (push–pull) systems which amplify the control current. With this configuration a diode can be driven by a current of several tens of microamperes.

8.3. Parallel and series connection

Figures 8.29 and 8.30 show two conventional circuits where the diode is connected either in parallel or in series. These two circuits operate in opposition; when one is illuminated, the other is extinguished.

8.4. Control of a diode with a TTL or CMOS device

Logic levels are available at the output of TTL and CMOS devices which can be fed to diodes. Care is necessary; a NAND gate such as the 7400 can accept 16 mA in the low state and thus easily feed a diode with a consumption of 10 mA. In contrast, a 74550 can output a maximum of only 8 mA, which is

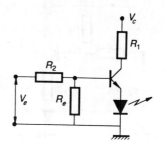

Figure 8.29   Series diode circuit

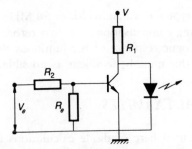

Figure 8.30 Parallel diode circuit

insufficient for the optimum efficiency of a conventional diode. In all cases it is preferable to use an open collector device such as the 7416.

For calculation purposes, the conventional equations established previously are again used.

8.5. Linear control of a diode

To transmit analogue signals on a fibre by means of a diode, it is essential to ensure a highly linear relationship between the control signal and the luminous flux. Figure 8.32 shows a basic circuit for driving a diode with a wideband video signal; it has an input impedance of 50 Ω.

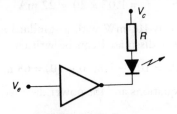

Figure 8.31 Basic diagram

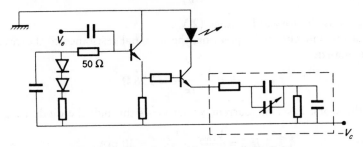

Figure 8.32 The principle of linear control of a diode

A system of this type has a bandwidth of 50 MHz which can be increased to 150 MHz by using compensation with two zeros. The diode is chosen by examining the analogue response of the luminous flux as a function of the excitation current; this must be as linear as possible.

9. TYPICAL EXAMPLES

Examples are presented here of diode calculations for a continuous and a pulsed case.

9.1. Example of a diode in continuous operation

A standard red diode is driven continuously to obtain a typical luminous intensity of 2.15 mcd at 25°C with an ambient temperature of 70°C. It will be driven by a TTL 7416 at 5 V (see Figure 8.31).

A standard diode is chosen which emits 2.0 mcd as a typical value at 20 mA. One has a nominal ratio:

$$\frac{I_{required}}{I_{nom}} = \frac{2.15 \text{ mcd}}{2.0 \text{ mcd}} = 1.07$$

The graph of relative luminous intensity gives

$$I_F = 1.07 \times 20 \approx 22 \text{ mA}$$

The power dissipated is 100 mW with a standard variation of 1.6 mW/°C above 50°C. The power dissipated can be written

$$P_m = 100 - 1.6(70 - 50) = 68 \text{ mW}$$

From the previous equations and the power dissipated it is possible to calculate the mean current:

$$I_m = \frac{P_m}{V_F} = 34 \text{ mA}$$

where V_F is the forward voltage (2 V).

By using conventional equations, the nominal resistance of the circuit can be calculated:

$$R = \frac{V_c - V_F}{I_F} = 136 \ \Omega$$

In the worst case, the current which flows through the diode is:

$$I_{max} = \frac{V_{cmax} - V_{Fmin}}{R_{min}} = 30 \text{ mA}$$

with:

$V_{cmax} = 5.5$ V
$V_{Fmin} = 1.8$ V
$R_{min} = 123$ Ω

This current is again compatible with the calculated maximum mean current.

9.2. Example of a diode in pulsed mode

A high-efficiency red diode, with a duty cycle of 10% at 1 kHz, is driven by an npn transistor (see Fig. 8.26). The required luminous intensity is 1.5 mcd at 25°C with a maximum operating temperature of 70°C.

A standard diode is chosen which emits 2 mcd using a supply of 5 V with a forward voltage of 2 V and a specified current of 10 mA. Knowing the duty cycle of illumination it is possible to calculate

$$I_p = \frac{\phi_m \cdot I_s \cdot \rho_c}{\tau \cdot r_p \cdot \phi_{L0}} = 50 \text{ mA}$$

with

$\phi_m = 1.5$ mcd	= required mean luminous intensity
$\phi_{L0} = 2$ mcd	= specific luminous intensity
$I_s = 10$ mA	= specific current
$\rho_c = 1$	= specific luminous efficiency
$\tau = 0.1$	= duty cycle
$r_p = 1.5$	= required efficiency

Knowing the peak current it is possible to calculate the mean current:

$$I_m = \tau \cdot I_p = 5 \text{ mA}$$

The power dissipated in the temperature range considered:

$$P = 100 - 1.6(70 - 50) = 68 \text{ mW}$$

Below 1 kHz, the duration of the pulse is 100 μs and the maximum tolerable ratio is 3 as shown in Fig. 8.21; in this case the maximum peak current is

$$I_{pmax} = 60 \text{ mA}$$

which is greater than the peak current calculated for the present case. Calculation of the nominal resistance uses conventional equations and gives

$$R = \frac{V_c - V_F}{I_p} = 60 \text{ Ω}$$

In the worst case, the maximum peak current which flows through the diode will be

$$I_{pmax} = 66 \text{ mA}$$

with

$$V_{cmax} = 5.5 \text{ V}$$
$$V_{Fmin} = 1.8 \text{ V}$$
$$R_{min} = 54 \text{ }\Omega$$

and in this case the diode operates at the limit of its capabilities.

10. THE BANDWIDTH OF A LIGHT-EMITTING DIODE

The bandwidth of the diode depends on the speed of radiative recombination in the material. The parameters which influence the duration of recombination are:

— the thickness (d) of the active region to be crossed, which affects the time to cross the junction;
— the doping of the active region (N_a), which arises with a low-injection diode;
— the density of the injected currrent (J_0), which depends on both the surface area of the active region and its distribution;
— the non-radiative recombination time (τ_{NR}).

These parameters are related by the following continuity equation:

$$\frac{dN}{dt} = \frac{J_0}{q \cdot d} - B \cdot N \cdot P - \frac{N}{\tau_{NR}} \tag{53}$$

where N is the electron density, P the hole density and B the spontaneous emission coefficient. The equation of neutrality gives

$$P = N_0 + N \tag{54}$$

In the steady state, electron density variation with time is zero:

$$\frac{dN}{dt} = 0 = \frac{J_0}{q \cdot d} - B \cdot N_0 \cdot (N_0 + N_a) - \frac{N_0}{\tau_{NR}} \tag{55}$$

from which the minority carrier density as a function of current density can be written

$$N(J_0) = \tfrac{1}{2} N_a' \cdot \left(\left(1 + \frac{4 J_0}{q \cdot d \cdot B \cdot N_a'^2} \right)^{1/2} - 1 \right) \tag{56}$$

with

$$N_a' = N_a + \frac{1}{B \cdot \tau_{NR}}$$

and the effective recombination time is

$$\tau(J_0) = \frac{q \cdot d}{2 \cdot J_0} \, N_a' \left(\left(1 + \frac{4J_0}{q \cdot d \cdot B \cdot N_a'^2}\right)^{1/2} - 1 \right) \tag{57}$$

Two diode families can be defined according to the density of the injected current. Those with low injection density have bandwidths less than 50 MHz and good operating linearity; those with high injection density have bandwidths which can reach 100 MHz but poor linearity and a dependence on the injection.

In the case of low-density injection, one has:

$$\frac{4J_0}{q \cdot d \cdot B \cdot N_a'} \ll 1 \tag{58}$$

hence

$$\tau = \frac{1}{B \cdot N_a'} \tag{59}$$

and the recombination time is independent of the current density.

In the case of high-density injection, one has

$$\frac{4J_0}{q \cdot d \cdot B \cdot N_a'} \gg 1 \tag{60}$$

hence

$$\tau(J_0) = \left(\frac{q \cdot d}{B \cdot J_0}\right)^{1/2} \tag{61}$$

and the recombination time depends on the current density and the thickness of the active region.

It is generally accepted that the response time of a diode (for the flux to rise from 10% to 90%) following a current pulse is approximately

$$t_r \simeq 2.2\tau \tag{62}$$

which enables the bandwidth of the diodes to be determined:

$$\Delta f \simeq (2 \cdot t_r)^{-1} \tag{63}$$

The bandwidth of the diode also depends on its control circuit and the bandwidth can be increased by ensuring a continuous reverse bias of the diode.

11. EXAMPLES OF LIGHT-EMITTING DIODES

Table 8.3 presents a comparison of three light-emitting diodes operating at the same wavelength ($\lambda = 860$ nm).

For diodes which are quite similar optically, there can be marked differences in response time, output power and power consumption. It is also necessary to consider performance improvements with respect to ambient temperature. The diodes presented in this table have maximum bandwidths of 40–70 MHz. In general, to ensure the maximum reliability of the link, the rise and fall times should be typically 5–10 times less than that of the signal. The bandwidths of these diodes are then much less (8–14 MHz).

Table 8.3 Comparison of three types of light-emitting diode

Type	Thomson (SG 12)	Honeywell SE 3362 04	H.P HFBR1204
Reverse voltage	1.8	1.8	1.7
Current (mA)	50	100	100
Optical power emitted (μW) in 50/125 fibre	35	25	12
Numerical aperture	0.20	0.21	0.38
Response time (ns)	7	12	11
Bandwidth (nm)	50	35	40

9

LASER DIODES

1. INTRODUCTION

The principle of the laser (light amplification by stimulated emission of radiation) is the combination of two essential elements:

— a light-wave amplifier;
— a feedback loop which forms a resonator.

The amplifier uses the properties of stimulated emission of a photon by an excited particle. Exploitation of this stimulated emission requires an important modification to the medium — the greatest possible inversion of its active population.

The resonator is an optical cavity in which the light wave is reflected and amplified. The simplest is the Fabry–Pérot resonator, which consists of two plane mirrors, one of which is semi-transparent. Other cavities which use more complex optics (such as a plane mirror and a concave mirror) exist but are not used in laser diodes.

2. THE AMPLIFYING MEDIUM

Consider a medium of which N_1 atoms are in energy state E_1 and N_2 in state E_2. The energy to pass from one state to the other is

$$E_g = E_2 - E_1 = hf$$

which represents the emission or absorption of a photon.

2.1. Absorption

Suppose that the material is in an electromagnetic field of frequency f; by absorbing energy hf, an electron can pass from state 1 to state 2 and, per unit time, one has

$$\frac{dN}{dt} = B_{12} \cdot N_1 \cdot \rho(f) \tag{1}$$

where $\rho(f)$ characterizes the spectral density of the radiation and N is the number of electrons making the transition.

2.2. Stimulated and spontaneous emission

Only state E_1 of the atom is stable and the electrons will have a tendency to return to this level; the time for which they remain there depends on their lifetime (τ) at this level:

$$\frac{dN}{dt} = A \cdot N_2 = \frac{N_2}{\tau} \quad \text{with} \quad A = 1/\tau \tag{2}$$

this is independent of the electromagnetic radiation and is spontaneous emission.

In contrast, the presence of radiation can also cause transitions from one level to another:

$$\frac{dN}{dt} = B_{21} \cdot N_2 \cdot \rho(f) \tag{3}$$

This stimulated emission is accompanied by emission of a photon, that is, an incident photon creates a second photon; this is the principle of laser amplification.

Furthermore, Planck has shown that, in a closed cavity, the spectral distribution is as follows:

$$\rho(f) = \frac{8\pi h f^3 n^3}{c^3} \cdot \exp\left(\frac{hf}{kT} - 1\right) \tag{4}$$

where n is the refractive index of the medium ($n = c/v$, v is the velocity of propagation of the wave in the medium), k is Boltzman's constant and T is the absolute temperature of the medium. One can write:

$$\frac{A}{B} = D(f) \cdot hf \tag{5}$$

where $D(f) = 8\pi f^2/c^3$ determines the number of vibratory states per unit volume. If $N(f)$ is the number of photons of frequency f, Planck's equation can be written:

$$\rho(f) = D(f) \cdot hf \cdot N(f) \tag{6}$$

which determines the spectral distribution as a function of the number of photons per vibratory state for a given frequency. It can be deduced that

$$B = \frac{c^3}{hf^3 n^3 8\pi} \cdot \frac{1}{\tau} \tag{7}$$

A and B are the Einstein constants.

2.3. The principle of the laser

As stimulated emission is accompanied by creation of a photon, it must be privileged; for this, a population inversion $(N_2 > N_1)$ is achieved by pumping.

If it is arranged that the lifetime τ is sufficiently long not to disturb the phenomenon, the induced emission is more powerful than that absorbed; this is the laser effect. The photon created in this way has the same direction, the same phase, the same polarization and the same frequency as the incident photon.

In thermal equilibrium, the population exchanges can be written

$$(N_1 - N_2)B \cdot \rho(f) + A \cdot N_2 = 0 \tag{8}$$

$$\rho(f) = \frac{AN_2}{B_{12}N_1 - B_{21}N_2} \tag{9}$$

$$\rho(f) = \frac{A}{B_{12}} \cdot \frac{1}{\left(\dfrac{B_{12}N_1}{B_{21}N_2} - 1\right)} \tag{10}$$

Boltzman statistics give

$$N_1 = C \cdot \exp(-E_1/kT)$$
$$N_2 = C \cdot \exp(-E_2/kT)$$
$$\frac{N_1}{N_2} = \exp\left[-\frac{E_1 - E_2}{kT}\right] = \exp\left(-\frac{hf}{kT}\right) \tag{11}$$

It is assumed that ρ must tend to infinity as the temperature tends to infinity;

hence

$$B_{12} = B_{21} = B$$

$$\rho(f) = \frac{A}{B} \cdot \frac{1}{\exp\left[\dfrac{hf}{kT} - 1\right]} \tag{12}$$

Let $\rho(f)$ be the number of photons at a given time. The cavity (of volume V) does not remain in equilibrium since there are always fluctuations created by thermal unbalances; these can be written

$$V \cdot \frac{\mathrm{d}\rho(f)}{\mathrm{d}t} = (N_1 - N_2)B \cdot \rho(f) + A \cdot N_2 \tag{13}$$

now

$$\rho(f) = D(f) \cdot hf \cdot N \tag{14}$$

$$\frac{\mathrm{d}N}{\mathrm{d}t} = (N_1 - N_2)W \cdot N + W \cdot N_2 \tag{15}$$

with

$$W = (V \cdot D(f)) \cdot \tau \cdot \Delta f)^{-1} \tag{16}$$

Equation (15) is the laser equation independent of losses.

Assume, as a first approximation, that N_1 and N_2 are independent of time. Equation (15) is easily solved and gives

$$N(f) = \frac{N_2}{N_1 - N_2} \cdot [\exp[(N_1 - N_2)Wt] - 1] \tag{17}$$

by considering that there are no photons at time $t = 0$.

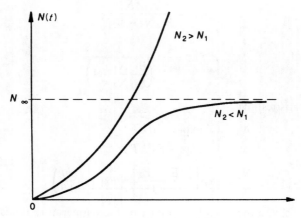

Figure 9.1 Variation of number of photons

Figure 9.1 shows the influence of population inversion on the occurrence of the laser effect.

3. THE RESONANT CAVITY

It is known that a resonant cavity permits amplification of a light-wave; it also permits selectivity of frequency, polarization, phase and the direction of the wave.

3.1. Design of the radiating cavity

In laser diodes, only Fabry–Pérot type cavities are used with a plane semi-transparent mirror to permit output of the beam.

Maxwell's equation can be written as follows for a perfect dielectric medium where E and H define the electric and magnetic fields:

$$\Delta E = \frac{1}{v^2} \cdot \frac{\partial^2 E}{\partial t^2}$$

$$\Delta H = \frac{1}{v^2} \cdot \frac{\partial H^2}{\partial t^2} \tag{18}$$

where v is the phase velocity of the wave in the medium. If ε is the permittivity and μ the permeability of the medium, one has

$$v = \frac{1}{(\varepsilon\mu)^{1/2}} \tag{19}$$

Assuming a plane wave, one obtains solutions of the form

$$E(x,t) = A \cdot \exp\left[i\omega\left(t - \frac{x}{v}\right)\right] + B \cdot \exp\left[-i\omega\left(t + \frac{x}{v}\right)\right] \tag{20}$$

Boundary conditions impose the following:

$$E(x_0, t) = E(x_1, t) = 0$$

$$A \cdot \exp\left[-i\omega\,\frac{x_0}{v}\right] = B \cdot \exp\left[-i\omega\,\frac{x_0}{v}\right]$$

$$A \cdot \exp\left[-i\omega\,\frac{x_1}{v}\right] = B \cdot \exp\left[-i\omega\,\frac{x_1}{v}\right] \tag{21}$$

The system must have a zero determinant in order to have non-trivial solutions:

$$\frac{2i\omega}{v}(x_1 - x_0) = 2\pi p \tag{22}$$

by putting

$$\omega = 2\pi f = \frac{2\pi\lambda}{v}$$

$$x_1 - x_0 = L \tag{23}$$

one obtains:

$$\lambda = \frac{2L}{p} \qquad p = 1, 2, 3, \ldots \tag{24}$$

Equation (24) is also called the phase condition.

Consequently, the presence of mirrors imposes a quantization of the frequencies which can propagate.

3.2. Separation of wavelengths

By differentiating the phase condition, one obtains

$$\Delta\lambda = \frac{2L}{p^2}\,\Delta p$$

for $\Delta p = 1$ one has

$$\Delta\lambda = \frac{2L}{p^2} \qquad \text{that is} \qquad \frac{\Delta\lambda}{\lambda} = \frac{1}{p}$$

Considering, for example, a cavity of length 10 cm centred for $\lambda = 6.328$ Å (the case of the He–Ne laser), one obtains

$$p = \frac{2L}{\lambda} = 316\ 055 \qquad \text{possible frequencies}$$

$$\Delta\lambda = \frac{\lambda}{p} = 0.02 \text{ Å} \qquad \text{between two modes}$$

Account must also be taken of the spreading of laser rays by the Doppler effect; this reduces the spatial resolution.

3.3. The lifetime of a photon

The lifetime of a photon in the cavity is related to the time which it requires to cross it; let this time be t_1:

$$t_1 = \frac{L}{v} \qquad (26)$$

If R is the reflection coefficient of the mirrors, the lifetime of the photon is

$$T_1 = \frac{t_1}{1 - R} = \frac{L}{(1 - R)v} \qquad (27)$$

The reflection R is never perfect, but in the case of a laser it is designed to allow a fraction of the beam energy to pass.

3.4. Reinforcement of modes in a Fabry–Pérot cavity

Consider the propagation of a wave in the cavity; after each reflection by the mirror, it is attenuated by a factor R and is subjected to a phase shift:

$$\phi = \frac{2\pi}{\lambda} \cdot 2L \qquad (28)$$

The total amplitude of the wave in the cavity can be written

$$a = A \cdot \exp(i\omega t) \cdot \sum_{n=0}^{n=\infty} R^n \exp(-i\phi n)$$

The luminous intensity can be written

$$I = a \cdot a^* = \frac{A^2 / (1 - R)^2}{1 + \frac{4R}{(1 - R)^2} \sin^2(\phi/2)} \qquad (29)$$

The initial intensity has a value I_0 and the maximum intensity is obtained for $\phi = k\pi$.

$$\frac{I_{max}}{I_0} = \frac{1}{(1 - R)^2} \qquad (30)$$

if $R = 0.9$, then $I_{max} = 100 \cdot I_0$.

Observe that for a high reflection coefficient of the faces, 100 times the initial intensity is obtained.

3.5. The fundamental laser equation

In the laser equation, no account has been taken of the lifetime of the photon

(T_1) in the cavity and the generalized laser equation can be written

$$\frac{dN}{dt} = (N_1 - N_2)W \cdot N + N_2 \cdot W - 2K \cdot N \qquad (31)$$

with

$$K = \frac{1-R}{L} \cdot v = \frac{1}{T} \qquad (32)$$

and

$$W = \frac{c^3}{V 8\pi f^2 \, \Delta f \tau}$$

The condition for laser radiation can then be deduced:

$$\frac{dN}{dt} > 0 \qquad (33)$$

by neglecting the spontaneous emission $(N_2 W)$, the condition for laser radiation is obtained:

$$\frac{(N_2 - N_1)v^3}{V 8\pi f^2 \, \Delta f \tau} > \frac{1}{T} \qquad (34)$$

Several points emerge from this radiation condition. Firstly, the lifetime of the photons must be as long as possible; cavity length can be increased. However, the volume must be as small as possible; hence a very small cross section is required. Finally, to increase the radiation, it is simplest to increase the population inversion.

Consider by way of example a laser cavity 30 cm long and of 1 cm^2 cross section which is required to resonate at 0.6328 μm. The separation between two frequencies is

$$\Delta f = \frac{c}{2L} = 500 \text{ MHz}$$

The population inversion in the cavity must be such that

$$N_2 - N_1 > \frac{1}{T} \cdot V \cdot \frac{8\pi f^2 \, \Delta f}{c^3} \cdot \tau$$

The reflection coefficient is 0.9 and the lifetime at the transition level is of the order of 10^{-3} seconds; hence

$$N_2 - N_1 > 35 \times 10^{10} \text{ atoms/cm}^3$$

To obtain the laser effect, pumping must ensure a population inversion much greater than 35×10^{10} atoms/cm^3 to obtain the power to compensate for the various losses.

The width of the spectral line and the lifetime τ in the excited level must observe Heisenberg's relation:

$$\Delta f \cdot \tau = 1 \tag{35}$$

The main causes of spreading of the laser channel are the Doppler effect, spreading of the energy levels, the Starck effect and inhomogeneity of the medium.

This greatly simplified theory of the laser effect in a cavity gives a good idea of the problems encountered and the parameters involved in analysing and realizing a laser diode.

4. LASER DIODES

4.1. Definition

Laser diodes are semiconductors in which an amplifying medium has been created together with a resonant cavity and in which population inversion is achieved by means of a current.

As long as the current remains below a threshold value, the laser diode behaves as a conventional light-emitting diode. As soon as the threshold is reached, population inversion is achieved and the laser effect is initiated.

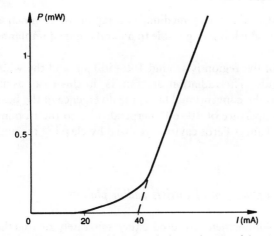

Figure 9.2   The power emitted by a laser diode as a function of the current

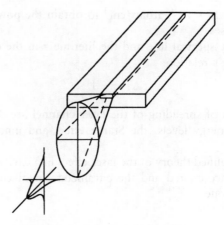

Figure 9.3 Basic diagram of a laser diode

Typically, the threshold current is of the order of 30–100 mA at 25°C but it varies and increases rapidly with temperature.

The threshold current depends to a large extent on the geometry of the cavity and particularly on the following:

— the length;
— the reflection coefficient;
— the thickness of the active region;
— the confinement factor which determines the manner in which the generated photons propagate in the cavity;
— the quantum efficiency.

In laser diodes, the active medium is a region of confinement of carriers and/or photons which is comparable to a parallelepiped implanted in the semiconductor.

The length of the region is around 100–500 μm and the width depends on the type of diode. The radiation diagram is, in this case, asymmetrical.

In contrast to the conventional laser, the divergence of the beam is large and amounts to an aperture of 30–60° perpendicular to the ribbon and 10° parallel to it. The Fabry–Pérot cavity is realized by cleaving the ends of the semiconductor.

4.2. Confinement of carriers and/or photons

Two types of confinement are used either separately or together in order to ensure the maximum efficiency:

— The active region of minority carriers is limited to their diffusion length which confines them to a small region and reduces the current required.
— Confinement of photons is obtained by doping the region surrounding the 'ribbon'; this modifies the refractive index of the medium and creates a waveguide.

4.3. Materials

The forbidden bands of the materials, which define the emitted wavelengths, depend on the material used. By modifying the proportions of the semiconductors used, its wavelength can be 'chosen'. The two main combinations used are

$$Al_xGa_{1-x}/As_ySb_{1-y}$$
$$Ga_xIn_{1-x}/As_yP_{1-y}$$

which allows conventional radiation to be obtained at 0.82 μm, 1.3 μm and 1.5 μm.

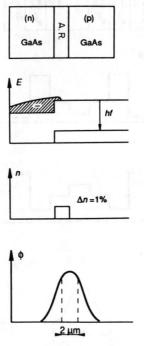

Figure 9.4 The principle of the homojunction

4.4. Structure

There are two types of structure — the homostructure and the heterostructure. The homostructure consists merely of a p–n junction with confinement of photons. The active region is of the same type as the p region but it is doped in such a way that its refractive index profile ensures confinement of photons.

In the case of the double heterostructure, there is a region of carrier confinement around the central region and then a region of optical confinement; this configuration enables the threshold current to be minimized.

w defines the thickness of the ribbon which is generally less than 1 μm. If the width of the ribbon is less than 15 μm, a resonance at the fundamental mode (TEM$_{00}$) is obtained, hence its use for fibre optics. As the width increases there is a change to a higher mode and the beam divides.

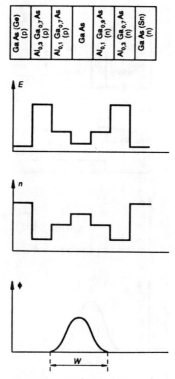

Figure 9.5 The principle of a heterostructure

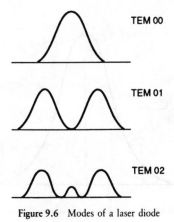

Figure 9.6 Modes of a laser diode

4.5. The emission spectrum

The Fabry–Pérot cavity resonates in several modes such as:

$$m\lambda = 2nL \tag{36}$$

where m is the order of the mode and n the refractive index of the medium. Differentiating this expression gives

$$m\ d\lambda + \lambda\ dm = 2L\ dn$$

hence

$$d\lambda = \frac{\lambda^2}{2nL} \cdot \frac{dm}{\frac{1}{n} \cdot \frac{dn}{d\lambda} - 1} \tag{37}$$

When L increases $\Delta\lambda$ decreases; since the length of the cavity of a laser diode is much shorter than for a conventional laser, the separation between modes is much greater. A simple numerical application with:

$$\lambda = 0.9\ \mu m$$
$$n = 3.6$$
$$L = 300\ \mu m$$

gives $\Delta\lambda = 0.4$ for the mode separation which is much greater than the mode separation of conventional lasers.

For a laser diode, the half intensity width is several nanometres for a peak wavelength of 820 nm. This is true only for a current greater than the threshold current; otherwise the configuration is that of a conventional diode.

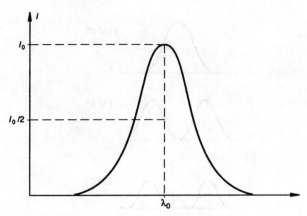

Figure 9.7 Width of a spectral line as a function of electric field strength

4.6. Thermal characteristics of laser diodes

The threshold current varies with temperature following an exponential law:

$$I_s = I_0 \cdot \exp(T/T_0) \tag{38}$$

Furthermore, heating of the active region occurs due to the Joule effect:

$$T - T_a = R_T \cdot V \cdot I \cdot r \tag{39}$$

where T_a is the ambient temperature, R_T the thermal resistance, V the voltage, I the current and r the pulse rate. This imposes a current on the diode given by

$$I = \frac{T - T_a}{R_T \cdot V \cdot r} \tag{40}$$

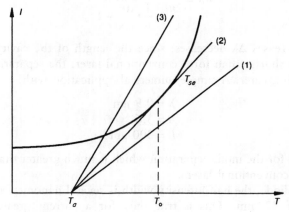

Figure 9.8 Current variation as a function of temperature

To operate, the diode must reach a·temperature which corresponds to the intersection of the curves of Equations (38) and (40). As a consequence, there will be cases where the laser diode can operate only in pulsed mode.

The threshold thermal resistance is defined as the single point of intersection of the two curves. In case (1), the two curves do not have a common point since $R_T > R_{Tth}$; there is, therefore, no laser effect. Case (2) is the limiting case where the laser effect can start and curve (3) represents a fully active laser effect.

4.7. Speed of response and modulation frequency

The simplest method is to modulate the current passing through the diode. When the threshold current is exceeded, a delay τ_d is observed in the modulation due to the lifetime τ_e of the minority carriers; it is this delay which determines the upper limit of the modulating frequency.

The current continuity equation can be written

$$\frac{dn}{dt} = \frac{I}{q \cdot A \cdot d} - \frac{n}{\tau_e} \tag{41}$$

where $A \cdot d$ is the volume of the active region and n is the number of minority carriers. The first term of the equation determines the uniform generation of minority carriers while the second determines the rate of recombination of these carriers. If it is assumed that initially no minority carriers exist ($n(0) = 0$), one obtains

$$n(t) = \frac{\tau_e \cdot I}{q \cdot A \cdot d} \left[1 - \exp\left(-\frac{t}{\tau_e} \right) \right] \tag{42}$$

At the threshold, the number of these carriers is put equal to n_{th} by defining a threshold current:

$$I_{th} = n_{th} \cdot q \, \frac{A \cdot d}{\tau_e} \tag{43}$$

which gives a bias delay:

$$t_d = \tau_e \cdot \log\left[\frac{I}{I - I_{th}} \right] \tag{44}$$

To reduce this time, the diode is pre-biased with a current I_0, that is initially $I = I_0 \neq 0$:

$$n(0) = \tau_e \cdot \frac{I_0}{q \cdot A \cdot d} \tag{45}$$

from which:

$$t_d = \tau_e \cdot \log\left[\frac{I - I_0}{I - I_{th}}\right] \tag{46}$$

Modulation is achieved by modulating the bias current; t_d then defines the modulation limit since the period of the signal must be greater than twice the minimum incremental period (Shannon's theorem).

The modulation limit then defines the bandwidth of the laser diode:

$$\Delta f = \frac{1}{2 \cdot t_d} \tag{47}$$

5. LASER DIODE OPTICAL HEADS

The optical cavity obtained by cleaving provides two active outputs. One will be coupled to the optical fibre leader, the other to the control photodiode which enables the power emitted by the diode to be controlled. Coupling of the optical fibre to the laser is usually made through matching optics. To reduce the temperature problem, the laser diode is generally mounted on a Peltier effect regulator.

5.1. Optical fibre–laser diode coupling

The coupling between the optical fibre and the laser diode must ensure that maximum power is injected into the optical fibre. The optical fibres generally used have a graded index with core diameters of 50 μm for multimode and 6 μm for monomode. This coupling is optimized by the presence of spherical

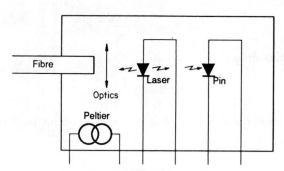

Figure 9.9   Configuration of a laser head

or hemispherical microlenses and limited by the accuracy of alignment. For this type of coupling, the losses are generally less than 4.5 dB.

5.2. Laser diode optical transmitter

The optical head cannot be used alone, it requires electronic control and protection circuits of a complexity which depends on the required performance. Four classes of function can be identified:

— protection of the laser diode;
— stabilization of the mean optical power;
— modulation of the laser current;
— control of modulation with a photodiode.

5.2.1. Protection of the diode

Three types of disturbance can be identified which can impair the operation of the diode:

— poor stability of the power supplies;
— interference on the power supplies;
— over-modulation of the diode.

It is thus necessary to stabilize and filter the laser power supplies and include protection against polarity reversal, over- and under-voltages.

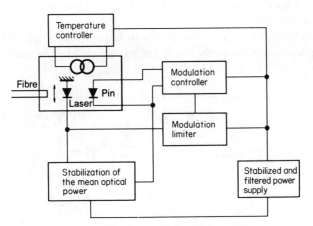

Figure 9.10 Configuration of a laser diode transmitter

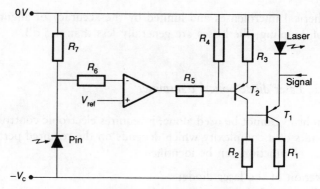

Figure 9.11   Basic principle of power stabilization

5.2.2. Stabilization of the mean optical power

This requires good power supply rejection and the use of lower than normal values for the currents and voltages.

In Fig. 9.11, the operational amplifier is driven by variation of current in the pin diode; it feeds transistor T_2, which in turn controls the current generator.

T_2 drives the base of T_1, which supplies power to the laser diode and controls the current which flows through it.

5.2.3. Modulation and modulation control

Modulation is achieved by variation of the current through the diode. Use of two back-to-back diodes limits the modulation; resistances R_1 and R_2 permit impedance matching.

This arrangement requires the diodes to have a very small series resistance and a capacitance sufficiently small not to cut off high frequencies.

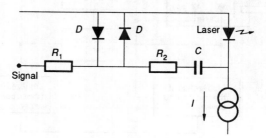

Figure 9.12   Modulation of a laser diode

6. LASER DIODE NOISE

Operation of a laser diode is accompanied by mode jumping and interference associated with the temporal coherence of the light wave; this leads to fluctuations of optical power at the receiver.

6.1. Mode jumping noise

According to its type (monomode or multimode), the laser light consists of one or more spectral lines. The wavelength emitted by the laser varies with the level of current injected and the temperature of the junction; the consequence is mode jumping, which generates non-linearities and noise at the receiver. The problem decreases as the number of modes increases.

6.2. Laser coherence noise

The laser emits a coherent light wave in the propagating medium. There will be inteference between two emitted light rays which follow different optical paths L_1 and L_2 to an observation point if

$$L_1 - L_2 = k\lambda$$

where k is an integer.

The presence of interference at the observation point is evidenced by variations of intensity in time and position of the observed spot, which depend on

— stresses applied to the fibre;
— the number of modes emitted and transmitted by the fibre;
— the current injected into the laser.

The variations are evidenced by noise at the receiver.

7. CHARACTERISTICS OF LASER DIODES

Table 9.1 presents the usual characteristics of commercial laser diodes. The example presented is a laser diode optical head from Thomson (Ref. SE 611).

Optical heads usually consist of a laser diode mounted on a heat dissipator, a Peltier effect module, a pin photodiode for control of the light emitted by the rear face of the laser diode, a leader fibre (in the present case a 50/125 μm multimode optical fibre) and a thermistor to control the temperature of the laser diode.

Table 9.1. An example of laser diode characteristics

Wavelength	1300 nm
Spectral width	2 nm
Threshold current	30 mA
Forward voltage	1.3 V
Output power on 50/125 multimode	3 mW
Response time	0.2 ns
Bandwidth	2.5 GHz
Sensitivity of the control photodiode	0.2 A/W
Response time of the control photodiode	3 ns

The particular advantage of laser diodes in comparison with light-emitting diodes is the wide bandwidth (typically gigahertz compared with 50 MHz for a light-emitting diode) and a high luminous power. The two main disadvantages are associated with power regulation and diode temperature.

10
THE RECEIVER

1. PHOTON–ELECTRON CONVERSION

1.1. The principle

The basic principle of reception of luminous flux by an electronic system is the conversion of photons into electrons.

The energy of the photon, in penetrating into the solid, is transferred to an electron in the valence band which then passes into the conduction band producing a pair of free carriers which generate a photocurrent.

The photocurrent created by the absorption of photons is proportional to the number of photons received. It is evident that the energy carried by the photon must be sufficient to permit a transition from the valence band to the conduction band; hence the forbidden bands should be made as narrow as possible:

$$hf \geqslant E_c - E_v = E_g \tag{1}$$

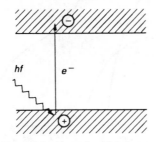

Figure 10.1   Absorption of a photon

Table 10.1

	As	Si	Ge	InSb
Gap (eV)	1.43	1.12	0.72	0.18
$\lambda_c(\mu m)$	0.87	1.1	1.7	6.3

the critical wavelength is

$$\lambda_c = \frac{hc}{E_g} \tag{2}$$

Due to thermal agitation, an electron which has left the valence band is subject to Brownian motion, having a mean velocity v_{th}, a relaxation time τ between two collisions and a mean free path l. In the case of germanium, for example, one has:

$$v_{th} = 10^5 \text{ m/s}$$
$$\tau = 1 \text{ ps}$$
$$l = 0.1 \ \mu m.$$

1.2. Absorption

Absorption of luminous flux in a material follows an exponential law:

$$\phi = \phi_0 \cdot \exp(-\alpha x) \tag{3}$$

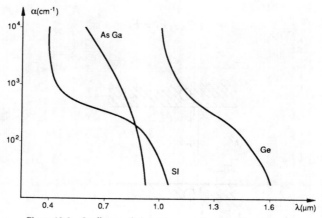

Figure 10.2 Coefficient of absorption as a function of wavelength

where α is the absorption coefficient, x the distance travelled in the material and ϕ_0 the initial flux.

In general, 95% of the incident radiation is absorbed over a distance of $x/3\alpha$. The coefficient of absorption depends strongly on the wavelength and the material used. By way of example, ultra-violet is absorbed in the first few microns of silicon while at least 100 μm are needed for red light.

In view of its response curve, silicon is the material used most for photo-detectors at 820 nm; germanium, in contrast, is used at 1.3 μm. These correspond to the two absorption windows of silica.

1.3. Quantum efficiency

Quantum efficiency ρ is the ratio of the number of electron–hole pairs created to the number of photons received:

$$\rho = \frac{I_{ph}/q}{\phi/hf} \qquad (4)$$

where

I_{ph} = detector photocurrent
q = charge of the electron
ϕ = incident flux
hf = energy of a photon.

Three principal factors reduce the quantum efficiency as follows:

— Losses by reflection at the input lens.
— Bulk recombination of created carriers. To reduce this one attempts to produce carrier pairs in the depletion region where there is an electric field which separates the generated pair; this reduces the transit time of the carriers and hence the probability of recombination.
— Surface recombination, which creates leakage currents. This effect is effectively combated by the use of guard rings.

1.4. Sensitivity

This is expressed in amperes/watt and follows from the quantum efficiency:

$$S_\lambda = \frac{I_{ph}}{\phi_0} \qquad (5)$$

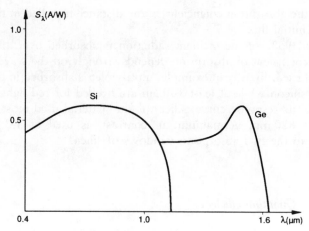

Figure 10.3   The spectral response of detectors

1.5 Spectral response

The spectral response determines the sensitivity of the receiver as a function of wavelength (Fig. 10.3).

The response of detectors depends on the material used; silicon is generally used for the visible region up to 1 μm and germanium for the near infra-red, notably 1.3 and 1.5 μm. Above this, compounds such as InSb and HgCdTe are used.

2. THE PRINCIPLE OF PHOTODETECTORS

Absorption of a photon by a semiconductor initiates the creation of an electron–hole pair which is separated; the electron goes to the n region and the hole to the p region.

Separation of the electron–hole pair is facilitated in regions where there is an electric field. The presence of an opposing electric field at the junction connections attracts electrons to the positive part and holes to the negative part; this creates a depletion region at the centre.

The depletion region exists even in the absence of an external electric field; it is due to the internal production of an electric field by the diffusion of minority carriers across the junction.

If the pair of carriers is generated outside the depletion region, it modifies the local concentration of free carriers and causes diffusion, which tends to make the concentrations uniform. When creation has occurred at a greater

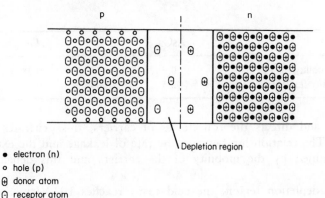

- • electron (n)
- ○ hole (p)
- ⊖ donor atom
- ⊖ receptor atom

Figure 10.4   The principle of a p–n junction

distance from the depletion region than the diffusion length, the probability that a recombination will cancel the pair of carriers is close to unity and the photon will have been absorbed without making a useful contribution to the current.

Let n_n be the number of electrons in the n region and n_p the number in the p region; the various concentrations can be written as follows (the subscript indicates the region):

$$n_p = n_n \cdot \exp\left(-\frac{qV_0}{KT}\right) \tag{6}$$

$$p_n = p_p \cdot \exp\left(-\frac{qV_0}{KT}\right) \tag{7}$$

Displacement of free carriers can be caused by an external electric field or a difference in concentration. The presence of an external electric field

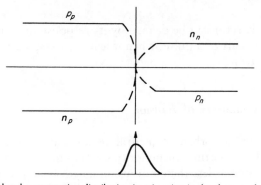

Figure 10.5   Field and concentration distribution in a junction in the absence of an external field

Table 10.2

	AsGa	Si	Ge	InSb
Electron mobility ($m^2/V \cdot s$)	0.9	0.12	0.38	0.7
Hole mobility ($m^2/V \cdot s$)	0.05	0.05	0.19	0.10

modifies and directs the trajectories of carriers; this generates a leakage current. The relationship between the rate of leakage and the external field is determined by the mobility of the carriers and this depends on the material.

In the depletion region, the field easily reaches 10^4 V/cm and the carriers can reach a speed of 10^7 cm/s.

Excessive concentration in a region leads to the appearance of a local electric field; a migration of carriers proportional to the potential gradient arises and generates a diffusion current.

The diffusion coefficient is related to the mobility as follows:

$$D = \frac{KT}{q} \cdot \mu \qquad (8)$$

similarly, the diffusion length of these carriers is related to their lifetime τ in the material:

$$l = (D \cdot \tau)^{1/2} \qquad (9)$$

The total current density is the sum of the direct and diffusion currents, which can be written:

$$J_T = J_d + J_{dif} \qquad (10)$$

3. SPEED

The speed of a detector can be defined by its response time; this specifies the time between the start of a pulse, when it leaves its rest state, and the instant when it returns to it.

3.1. The influence of diffusion

When photons are absorbed in neutral regions, the speed of response is limited by the diffusion time of minority carriers to the depletion region.

Only pairs created at less than a diffusion length from the depletion region contribute to the signal; the others are absorbed by the medium. If the useful thickness to be crossed is equal to the diffusion length, the diffusion time is

$$t_{dif} = \frac{1}{V_{dif}}$$
(11)

The effect of diffusion is reduced if the junction is formed near the surface and it operates in a region of sufficiently large space charge.

3.2. The influence of transit time

In the depletion region, the electric field is of the order of 10^4 V/cm and the carriers rapidly attain their limiting drift velocity, which is of the order of 10^7 cm/s. If w is the width of the depletion region, the transit time to cross this region can be written

$$t_{trans} = w/V$$
(12)

3.3. Junction speed

Transit and diffusion times constitute the intrinsic response time of the photodiode since they depend on neither the surface of the component nor the circuit in which it is used; currently 1 ps can be achieved.

Due to the presence of a p carrier region next to an n carrier region, the junction is equivalent to a dipole of charge whose capacitance varies with the bias voltage and the width of the depletion region; a typical value is $1-2$ μF. This capacitance directly limits the bandwidth of the diode, which is increased by increasing the width of the junction.

4. THE PIN PHOTODIODE

4.1. The principle

The pin photodiode is a conventional photodiode in which a region of high resistivity (the intrinsic region I) has been inserted between the two carrier regions; the quantity of free carriers is reduced by operating under a voltage.

When the junction is reversed biased, the depletion region increases and the majority carriers are incapable of crossing it; the only current (I_{th}), called the threshold, which exists is due to the passage of carriers which were initially minority.

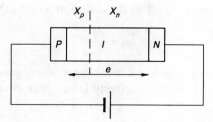

Figure 10.6   The principle of a pin photodiode

4.2. Current calculation

The rate of generation of pairs, assuming the absence of thermal current and the thickness of the surface to be equal to the diffusion length, is given by

$$G(x) = \phi_0 \cdot \alpha \cdot \exp(-\alpha x) \tag{13}$$

the current from the diode can then be written

$$J_d = q \cdot \int_0^1 G(x) \cdot dx$$
$$J_d = q \cdot \phi_0 [1 - \exp(-\alpha l)] \tag{14}$$

For $x > 1$, the minority carrier density n_p is obtained from the equation

$$D_n \frac{d^2 n_p}{dx^2} - \frac{n_p - n_{p0}}{\tau_n} + G(x) = 0 \tag{15}$$

where:

D_n = diffusion constant
τ_n = lifetime of the n carriers
n_{p0} = equilibrium density of the n carriers

from which can be obtained

$$n_p = n_{p0} + \phi_0 \cdot \frac{\alpha \cdot \tau_n}{1 - \alpha^2 D_n \cdot \tau_n} \cdot \exp(-\alpha \dot{x}) \tag{16}$$

and the diffusion current has a value

$$J_{dif} = -q \cdot D_n \cdot \frac{dn_p}{dx} \bigg|_{x=l} \tag{17}$$

$$J_{dif} = q \cdot \phi_0 \cdot \left[\frac{\alpha^2 \cdot \tau_n \cdot D_n}{1 - \alpha^2 D_n \cdot \tau_n} \right] \cdot \exp(-\alpha l) \tag{18}$$

the total current can be written

$$J_T = q \cdot \phi_0 \left[1 - \frac{\exp(-\alpha l)}{1 - \alpha^2 D_n \cdot \tau_n} \right] \tag{19}$$

To obtain the maximum current, it is necessary that

$$\alpha l \gg 1$$
$$\alpha^2 D_n \cdot \tau_n \gg 1$$

which is opposite to the speed condition; the usual compromise is $l = 1/\alpha$.

4.3. Dark current

In the absence of all external light, a current always flows in the diode; this is due to the terminal voltage which causes minority carriers to diffuse across the junction. This dark current can be written

$$I_0 = I_{th} \cdot \left[\exp\left(\frac{qV_d}{kT}\right) - 1 \right] \tag{20}$$

where V_d is the reverse bias voltage of the diode ($V_d < 0$) and I_{th} the threshold current under reverse bias. This current increases as the temperature increases.

4.4. Equivalent circuit of the pin diode

$$I = I_0 + I_{ph} \tag{21}$$

$$I_{ph} = S\lambda \cdot \phi_0 \tag{22}$$

In general, the current flowing in the diode is the sum of the dark current and the photocurrent. The dark current has typical values less than 1 nA; the internal resistance of the diode is of the order of 10^{11} Ω.

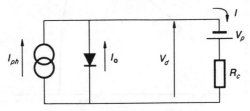

Figure 10.7   Equivalent circuit of a pin diode

5. STRUCTURE

The basic structure is similar for both conventional photodiodes and pin photodiodes; only the definition of the depletion region differs.

The pin diode is a chip of length 1250 μm and thickness 50 μm with the following distinct regions:

— an aluminium cathode;
— a doped semiconductor n region;
— an intrinsic semiconductor region;
— a p region ≈ 0.5 μm to limit diffusion outside the depletion region;
— a sensitive circular region with a diameter of 1000 μm covered with an anti-reflectant and surrounded by an 8 μm layer of silicon on which the aluminium anode is deposited.

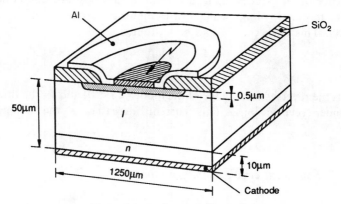

Figure 10.8   Configuration of a pin diode

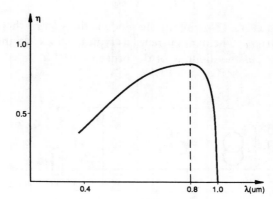

Figure 10.9   Quantum efficiency of a pin diode

The optimum efficiency of this type of configuration is of the order of 0.7 centred on 820 nm for a silicon diode.

6. THE AVALANCHE PHOTODIODE (APD)

6.1. The principle

Under the action of a sufficiently strong electric field ($V > 10^5$ V/m), a carrier can reach the ionization threshold of the material and create an electron–hole pair by collision with an atom of the lattice. If this in turn reaches the threshold, it creates other carriers and so on. The phenomenon becomes cumulative and conduction occurs by avalanche; this is the avalanche principle.

The avalanche gain depends strongly on the electric field and the variation in carrier ionization coefficient; these are determined by the number of electron–hole pairs created by a carrier in unit distance. Since each ionization is a random event, the instantaneous gain m fluctuates very rapidly and introduces excess noise defined by a noise factor particular to the avalanche diode.

To realize such avalanche photodiodes having a high gain–bandwidth product and a low excess noise factor, it is necessary to use materials whose ionization coefficients are as different as possible and to realize a structure such that the avalanche is initiated by carriers having the highest ionization coefficient.

6.2. Avalanche gain

The mean avalanche gain can be defined as the ratio of the current after multiplication to the current before multiplication:

$$M = \langle m \rangle = \frac{I - I_{m0}}{I_p - I_0} \tag{23}$$

where:

I   = primary current before avalanche
I_0 = dark current
I_{m0} = dark current with avalanche
I = total measured current.

An empirical relation gives a value for the multiplication factor:

$$M = \frac{1}{1 - \left(\dfrac{V_r - I \cdot R}{V_b}\right)^n} \tag{24}$$

V_b = breakdown voltage
R = series resistance of the diode
V_r = reverse voltage
n = a factor depending on the doping.

The maximum is attained when $V_r = V_b$ with $I \cdot R \ll V_b$ and can be written:

$$M_{max} = \frac{V_s}{n \cdot I \cdot R} \tag{25}$$

As a first approximation, the dark currents are neglected; in this case, the photocurrent of an avalanche photodiode is

$$I_{ph} = S_\lambda \cdot \phi_0 \cdot M \cdot m \cdot (2)^{-1/2}$$

and this reduces to a conventional photodiode for $M = 1$.

6.3. The structure of an avalanche photodiode

In this type of photodiode using high electric fields, there are two points to be considered — the electric field must be homogeneously distributed and the

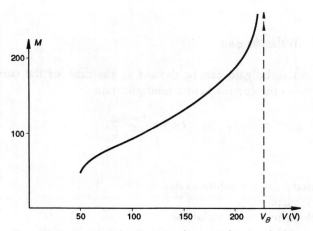

Figure 10.10   Variation of multiplication factor as a function of voltage

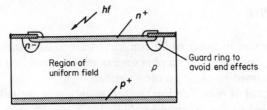

Figure 10.11 Configuration of an avalanche photodiode

material must be as homogeneous as possible. In general, the creation of regions which can cause breakdown effects in the diode must be avoided.

7. EQUIVALENT ELECTRONIC CIRCUIT OF A PIN DIODE

It is known that the photocurrent is proportional to the luminous flux (stray or signal). The pin photodiode behaves as an analogue component; its characteristics therefore depend on the preamplifier stage which follows it.

From an electronic point of view, a photodiode is characterized by:

— the series resistance (R_s), which is very small and generally negligible in comparison with the junction resistance (R_d);
— the junction capacitance (C_d);
— the dark current (I_0);
— the thermal current (I_{th});
— the quantum current (I_q);
— the photocurrent (I_{ph}).

The input preamplifier can be characterized by its input impedance $(R_a$ and $C_a)$ which enables the impedance of the system to be defined:

$$R_c = \frac{R_a \cdot R_d}{R_a + R_d} \tag{26}$$

$$C_c = C_d + C_a \tag{27}$$

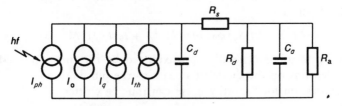

Figure 10.12 Equivalent circuit of a pin photodiode

8. NOISE

The various sources of noise to be considered are mainly thermal noise and quantum noise; both are random phenomena which depend on the bandwidth (Δf) and are defined by their quadratic mean.

In this definition of noise (and in the following), account will not be taken of noise generated by the preamplifier (whether a current source or a voltage source) but only that related to its impedance.

8.1. Thermal noise

This noise, associated with thermal agitation, appears in every conductor regardless of its nature. The expression for a load resistance R_L is

$$\langle I_{th}^2 \rangle = \frac{4KT}{R_L} \cdot \Delta f \tag{28}$$

8.2. Quantum or Schottky noise

Quantum or Schottky noise arises from the random distribution of the incident photons; the simplified expression is

$$\langle I_s^2 \rangle = 2q \cdot \Delta f (I_{ph} + I_0) \tag{29}$$

which signifies that the noise arises in the receiver as well as the received signal.

The photocurrent includes the current generated by the optical flux incident on the fibre and stray background light. Minimum Schottky noise is thus obtained for $I_{ph} = 0$, I_0 being the dark current.

8.3. Avalanche diode noise

Using the same reasoning but taking account of the noise factor $F(M)$ due to the creation of electron–hole pairs, the expression for Schottky noise for avalanche diodes is

$$\langle I_s^2 \rangle = 2q(S_\lambda \cdot \phi_0 \cdot m \cdot (2)^{-1/2} + I_0)F(M)M^2 \cdot \Delta f \tag{30}$$

The noise factor is generally approximated by the function $F(M) = M^x$ where x depends on the diode (it reverts to the pin diode for $M = 1$). The general expression for Schottky noise can thus be written

$$\langle I_s \rangle^2 = 2q(I_{ph} + I_0) \cdot M^{2+x} \cdot \Delta f$$

9. SIGNAL-TO-NOISE RATIO

9.1. Signal-to-noise ratio

As its name indicates, the signal-to-noise ratio defines the magnitude of the signal with respect to the noise; in general, the signal power must not be less than the noise power to ensure satisfactory detection.

$$\frac{S}{N} = \frac{R_L \cdot \langle I_{ph}^2 \rangle}{R_L [\langle I_s^2 \rangle + \langle I_t^2 \rangle]} \tag{31}$$

The incident optical power can be 'all or nothing' or modulated about a mean value; the expression is

$$\phi = \phi_0 [1 + m \cdot \exp(i\omega t)] \tag{32}$$

where m is the modulation depth. Let I_{ph} be the current determined by the effective power of the optical flux for any photodiode (pin or avalanche); one has

$$\langle I_{ph} \rangle = S_\lambda \cdot \phi_0 \frac{m}{2^{1/2}} \cdot M \tag{33}$$

the general expression for signal-to-noise ratio is

$$\frac{S}{N} = \frac{\phi_0^2 \cdot S_\lambda^2 \cdot m^2/2 \cdot M^2}{2q(\langle I_{ph} \rangle + \langle I_0 \rangle)M \cdot \Delta f + \dfrac{4KT}{R_L} \Delta f}$$

hence, substituting for $\langle I_{ph} \rangle$:

$$\frac{S}{N} = \frac{\phi_0^2 \cdot S_\lambda^2 \cdot m^2/2}{2q \cdot \Delta f \left(S_\lambda \phi_0 \dfrac{m}{2} + I_0 \right) M^{2+x} + \dfrac{4KT}{R_L} \Delta f} \tag{34}$$

9.2. Equivalent noise power (ENP)

The ENP determines the minimum input power to distinguish the signal from the noise, that is:

$$\frac{S}{N} = 1 \tag{35}$$

The minimum optical power ϕ_{0min} necessary to obtain a given signal-to-noise ratio is:

$$\phi_0 = (2)^{1/2} \cdot q \cdot \left(\frac{S}{N} \right) \cdot \frac{M^x \cdot \Delta f}{S_\lambda \cdot m} \cdot \left[1 + \left(1 + \frac{I_{eq}}{M^{2+2x} \cdot \Delta f \cdot q^2 \cdot (S/N)} \right)^{1/2} \right] \tag{36}$$

with

$$I_{eq} = 2q \cdot I_0 \cdot M^{2+x} + \frac{4kT}{R_c \cdot q} \tag{37}$$

It can be assumed that

$$\left[\frac{I_{eq}}{M^{2+2x}q^2(S/N)\,\Delta f}\right]^{1/2} \gg 1 \tag{38}$$

from which

$$\phi_{0min} = \frac{1}{S_\lambda \cdot m \cdot M} \cdot [2 \cdot I_{eq} \cdot (S/N) \cdot \Delta f]^{1/2} \tag{39}$$

It can be observed that the minimum optical power is inversely proportional to the sensitivity and the avalanche factor; this is to be expected and clearly shows the usefulness of avalanche diodes.

The ENP is thus defined by the expression for the effective optical power to obtain a signal-to-noise ratio of 1 for a bandwidth of 1 Hz:

$$ENP = \frac{\phi_0}{[2\,\Delta f(S/N)]^{1/2}} \quad \text{in} \quad W/(Hz)^{1/2} \tag{40}$$

$$ENP = \frac{1}{S_\lambda \cdot m \cdot M} \cdot [2I_{eq}]^{1/2} \quad \text{in} \quad W/(Hz)^{1/2} \tag{41}$$

where S_λ is the sensitivity at the specified wavelength. In principle, account should also be taken of optical noise associated with the fibre/diode interface; this generates a photocurrent which is added to the equivalent noise current.

9.3. Detectivity

Detectivity, in contrast to ENP, takes account of the surface area of the detector:

$$D = \frac{A^{1/2}}{ENP} \quad \text{in} \quad Hz^{1/2}\,m/W \tag{42}$$

10. ADDITIONAL CHARACTERISTICS

In their documentation manufacturers provide information in addition to the parameters described previously.

This includes, in particular, the type of package and whether or not it is equipped with a lens. Knowledge of the package type enables the type of fibre optic mounting which suits the photodiode to be determined.

In general, the indicated ENP is determined for Schottky noise and needs to be refined for a particular application.

The manufacturer also indicates the type of fibre for which the input optics have been optimized.

A final important parameter also provided is the variation of the various diode constants as a function of temperature.

11. SPECIFIC CIRCUITS

11.1. Photodiode and integrated preamplifier

In order to reduce the problems of electronic noise and stray capacitance, many diodes are now equipped with a preamplifier which provides a digital or analogue signal according to the application for which it has been designed. Their performance is lower than that of photodiodes alone but they are much easier to use; the difference is currently between several hundreds of megahertz for diodes and 5–10 MHz for the integrated devices.

11.2. Phototransistors

These are transistors whose base current is provided by the luminous flux.

Photodarlington devices are also available but are not used in optical telecommunications.

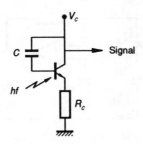

Figure 10.13   Basic diagram of a phototransistor

12. EXAMPLE CIRCUITS

The following two circuits are the basic configurations for using photodiodes with optical fibres. An example of receiver design is given in Chapter 11.

12.1. Photodiode with current amplifier

One of the simplest circuits is the photodiode followed by a preamplifier with negative feedback; this ensures approximately linear operation of the diode.

The advantage of the reverse bias voltage is to increase the speed of response of the diode but an amplifier with high input impedance is necessary in order to have a high open loop gain.

The diode is characterized by its load impedance (R_L, C_L); the feedback impedance of this circuit (R_f, C_f) enables the bandwidth of the circuit to be defined:

$$\Delta f = \frac{1}{2\pi \cdot R_f \cdot C_f}$$

12.2. Logarithmic circuit

In this case, the photon current is equal to the photocurrent:

$$I_p = I_{ph} \tag{43}$$

hence

$$I_p = I_s \cdot \left[\exp\left(\frac{V}{kT}\right) - 1 \right] \tag{44}$$

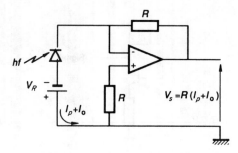

Figure 10.14 Linear photodiode circuit

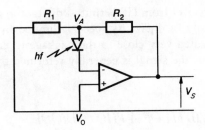

Figure 10.15   Logarithmic circuit

with

$$V = V_A - V_0$$

from which

$$V = \frac{kT}{q} \cdot \ln\left(1 + \frac{I_p}{I_s}\right) = \frac{R_1 \cdot V_s}{R_1 + R_2} \tag{45}$$

The expression for the output voltage is

$$V_s = \left(1 + \frac{R_1}{R_2}\right) \cdot \frac{kT}{q} \cdot \ln\left(1 + \frac{I_p}{I_s}\right) \tag{46}$$

The advantage of this circuit is the provision of a signal representing the logarithm of the incident flux.

12.3. Diode circuit equipped with a preamplifier

The main problem of this type of circuit is random noise whose amplitude has a square law variation; reduction of this noise allows the sensitivity of the system for signal acquisition to be improved and hence the bandwidth to be increased.

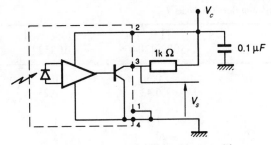

Figure 10.16   Circuit of a diode with preamplifier

The HFBR 2201 receiver from HP permits a digital throughput of 2.5 MHz with an input threshold of 4 μW; its speed is limited by noise. The 0.1 μF capacitor must be located very close to the package to ensure good filtering of power supply noise; the signal is generally reformed by a Schmitt trigger.

13. BANDWIDTH OF A PHOTODIODE

Two parameters impede the conversion of a received light signal into an electrical signal. These are, as has been seen, too low a signal-to-noise ratio, which prevents recognition of the signal, and inadequate receiver response time with respect to the transmitted bandwidth. The response time of the diode depends on its internal characteristics (related to the photon-electron conversion time) and its electronic interface circuit together with its internal capacity (several pF). Usual values of the bandwidth of a pin diode are from 100 to 500 MHz.

14. PIN DIODE CHARACTERISTICS

The critical parameters of a photodiode are normally the response time and the internal capacitance, which limit the bandwidth, the sensitivity and the noise level which increases the error rate. Table 10.3 gives three examples of pin photodiodes for a wavelength of 820 nm.

Observe that for identical response times, the sensitivity can vary by a factor of 2 and the dark current by a factor of 10. The choice of diode is made in accordance with the type of application.

For avalanche diodes (APD) in the same wavelength ranges, the usual parameters are given in Table 10.4.

Table 10.3 Comparison of three pin diodes

Type	H.P. 5082-4204	Honeywell 503322	Thomson SR 40
Dark current (nA)	0.6	0.1	1
Sensitivity	0.43	0.25	0.5
Junction capacity (pF)	2	2	2
Response time (ns)	1.5	1	1
Bias voltage (V)	− 20	15	20

Table 10.4   The parameters of an avalanche diode

Avalanche voltage	< 200 V
Capacitance	< 1 pF
Response time	< 1 ns
Multiplication coefficient	50
Sensitivity ($M = 1$)	0.5 A/W
Dark current	1 nA ($M = 50$)
Wavelength	820 nm

The performance is identical to that of a conventional pin diode as far as the response is concerned but the sensitivity due to the avalanche effect is 50 times better. However, there is increased avalanche noise which can limit the performance of the system.

11
APPLICATIONS OF OPTICAL FIBRES

As it is not possible to treat all the applications of optical fibres in this work, three application areas have been chosen which will be illustrated with practical examples. These are telecommunications, transducers and imaging.

1. OPTICAL FIBRE TELECOMMUNICATIONS

There are two main modes of telecommunication which are compatiable with optical fibres. These are digital transmission, intended particularly for transmission between computers, and analogue transmission intended for video transmission.

The main advantages of optical fibre transmission in comparison with conventional methods are:

— low attenuation which permits the use of long distances without repeaters;
— high bandwidth which can be increased by using frequency and/or wavelength multiplexing techniques;
— insensitivity to electromagnetic noise;
— the absence of radiation which minimizes the chance of eavesdropping;
— reduced cable length;
— low weight;
— excellent electrical insulation.

Transmission can be on a simple link or through a network; in both cases, performance evaluation involves analysis of elementary links.

1.1. Description of a fibre optic link

The principle of all data transmission is to send information from a transmitter to a receiver while minimizing the chance of distortion of the received signal and ensuring the maximum reliability of information transfer.

Figure 11.1 shows a conventional transmission chain in which data transfer is supported by optical fibre.

The signal is first coded or modulated in accordance with a known sequence which can be monitored at the receiver. This signal is injected into the optical fibre by the transmitter.

At the fibre output, the signal is received by a photodiode and then amplified before being decoded to return it to its original form.

Design of fibre optic links, therefore, consists of evaluating the degradation suffered by the reconstituted signal and choosing the best components to minimize this degradation.

1.2. Design of fibre optic links

In this section, the calculations necessary for the design of fibre optic links will be described with emphasis on the optical aspects, which include:

— the transmitter (diode or laser);
— the fibre;
— the receiver (pin or APD).

The interface characteristics between each of these will be analysed.

Every analysis of a fibre link depends on the previous chapters, particularly those concerning diodes, receivers and connectors, where all the equations used here have been explained in detail.

The flux which must be injected into the optical fibre depends on the noise

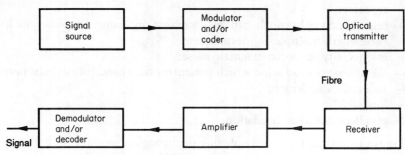

Figure 11.1   Fibre optic transmission chain

characteristics of the receiver, the bandwidth, the connector losses and the transmission losses in the cable. The flux at the receiver input must be sufficient for the signal-to-noise ratio of the received signal to exceed a specified minimum error rate.

Let ϕ_t be the flux available at the transmitter output and ϕ_r that available at the receiver input. Let L_{tf}, L_{fr} and L_{ff} be the transmitter–fibre, fibre–receiver and fibre–fibre losses respectively and L_{f0} that associated with the fibre alone. An equation for loss evaluation is obtained:

$$10 \cdot \log\left(\frac{\phi_r}{\phi_t}\right) = L_{tf} + L_{f0} + n \cdot L_{ff} + L_{fr} + M \qquad (1)$$

where n is the number of fibre–fibre couplings and M the required margin.

1.2.1. Losses at the transmitter–fibre interface

Losses at the transmitter–fibre interface are of three kinds:

(a) Losses associated with matching of the numerical apertures:

$$L_{NA} = 20 \cdot \log\left(\frac{NA_t}{NA_f}\right) \qquad (2)$$

where NA_t is the numerical aperture of the transmitter and NA_f that of the fibre.

(b) Losses associated with the ratio of the surface areas of the source and the fibre:

$$L_s = 20 \cdot \log\left(\frac{S_t}{S_f}\right) \qquad (3)$$

where S_t is the diameter of the transmitter and S_f that of the fibre core.

(c) Losses associated with mismatching of the refractive index gradients:

$$Lg = 10 \cdot \log\left(\frac{1 + 2/\alpha_f}{1 + 2/\alpha_t}\right) \qquad (4)$$

where α_f is the refractive index parameter of the fibre and α_t the refractive index parameter of the transmitter. This type of loss arises only in the case of transmitters with leader fibres.

(d) Fresnel losses associated with the air–silica interface:

$$L_F = 10 \cdot \log\left(\frac{4}{2 + n_f + n_f^{-1}} \middle| \frac{4}{2 + n_t + n_t^{-1}}\right) \qquad (5)$$

where n_t is the refractive index of the transmitter and n_f that of the fibre.

Usually, the flux defined for the transmitter takes account of the transmitter–air interface and the losses reduce to those of the air–fibre interface. As the fibre has a typical refractive index of 1.5, the Fresnel losses amount to 0.17 dB.

The overall assessment of losses at the transmitter–fibre interface can be written

$$L_{tf} = L_{NA} + L_g + L_F + L_S \qquad (6)$$

1.2.2. Losses in the fibre

These are caused by dispersion in the fibre and are expressed in dB/km; their amplitudes depend on the type of fibre used.

1.2.3. Losses at fibre–fibre interfaces

At a fibre–fibre interface, losses arise which are associated with the connection system itself and with the source-fibre/receiving-fibre interface.

If the two fibres are identical, only Fresnel losses arise; otherwise numerical aperture, refractive index gradient and surface area losses must be included.

The overall assessment at the fibre–fibre interface can be written

$$L_{ff} = L_{NA} + 2 \cdot L_F + L_g + L_S + L_{con} \qquad (7)$$

1.2.4. Losses at the fibre–receiver interface

The same types of loss as at the transmitter–fibre interface occur again but there is never a leader fibre; the overall assessment is

$$L_{fr} = L_{NA} + L_S + L_F \qquad (8)$$

Receivers generally have a large surface area and a large numerical aperture, in which case the losses reduce mainly to Fresnel losses.

$$L_{fr} \approx L_F \qquad (9)$$

Here again, the Fresnel losses are estimated at 0.17 dB.

1.2.5. The margin

The margin is an additional fictitious loss which enables an error tolerance to be included in the link. It must take account of ageing of the transmitter and be sufficient to compensate for this deterioration. Furthermore, it must not be too large in order to avoid exceeding the dynamic range of the receiver.

1.3. Design example

This will be the design of a 21 km fibre optic link without repeater having a bandwidth of 80 MHz and an error rate less than 10^{-9}. This type of design is performed in four stages in order to obtain:

— a specification for the detector;
— a specification of the link;
— a specification of the transmitter;
— an evaluation of the link.

1.3.1. Specification of the detector

The bandwidth of 80 MHz is not very large; a pin photodiode will be chosen which has a risetime of 300 ns without bias and 1.5 ns with a reverse bias of 20 V (for example an HFBR-4205 receiver).

For satisfactory transmission, the bandwidth of the receiver must be five times greater than the bandwidth of the received signal; this corresponds to rising and falling edges 10 times faster than the received signal. The bandwidth of the receiver is

$$Br = \frac{1}{5 \times 1.5^{-9}} \tag{10}$$

that is 133 MHz, which is much greater than the required 80 MHz.

For a load resistance greater than 10^{10} Ω, the thermal noise is negligible and hence the ENP depends only on the dark current:

$$\text{ENP} = \frac{(2qI_0)^{1/2}}{S_\lambda} \tag{11}$$

where q is the charge on the electron (1.6×10^{-19} eV) and S_λ the sensitivity at $\lambda = 820$ nm, the optimum wavelength of the detector. The numerical values are

$$S_\lambda = 0.43 \text{ A/W} \qquad I_0 = 0.15 \times 10^{-9} \text{ A} \tag{12}$$

hence

$$\text{ENP} = 1.6 \times 10^{-14} \ \mu\text{W/Hz}^{1/2} \tag{13}$$

To obtain an error rate less than 10^{-9}, a signal-to-noise ratio of 12 is necessary; hence an additional 11 dB is required. The sensitivity of the detector in dBm is

$$\text{ENP} = 10 \cdot \log\left(\frac{\text{ENP}}{1000}\right) \tag{14}$$

The sensitivity of this detector is -108 dB. Hence by taking account of the error rate, the following sensitivity is obtained:

$$S_{disp} = -108 + 11 = -97 \text{ dBm}$$
$$= 2 \times 10^{-7} \mu\text{W} \tag{15}$$

1.3.2. Specification of the link

The maximum length of the link is 21 km. A conventional optical fibre is chosen:

— multimode 50/125;
— 3.5 dB/km at $\lambda = 820$ nm;
— NA = 0.19;
— refractive index parameter = 2;
— bandwidth = 400 MHz/km;
— section length = 2.2 km.

The bandwidth is sufficient for this application and 10 sections are required to cover the distance. Coupling with bonded splices is chosen, giving 0.5 dB of loss. The loss of the link is

9 couplings	$0.5 \times 9 =$	4.5 dB
Fibre	$10 \times 2.2 \times 3.5 = 77$	dB
Total		81.5 dB

1.3.3. Transmitter specification

The transmitter must combine power and speed; a laser diode is chosen with the following principal characteristics:

— $\lambda = 840$ nm;
— $P = 3$ mW at the output of the leader fibre;
— response time of 1 ns;
— leader fibre:
  50/125 μm,
  multimode,
  NA = 0.23,
  refractive index parameter = 2.

The response time of 1 ns provides a bandwidth of 500 MHz; since 80 MHz is required, this is exceeded by a factor of 6.

The coupling between the optical fibre and the leader fibre is made with a connector (3 dB). The loss at the transmitter/fibre interface is as follows:

— connector   3   dB
— L_{NA} 1.7   dB
— L_f $2 \times 0.17 = 0.34$ dB
— L_g 0
— L_s 0
— Total ≈ 5.1 dB

1.3.4. Assessment of the link

This involves checking that the power emitted by the transmitter will arrive within the useful sensitivity of the detector. For this, one starts by assessing the losses:

— transmitter/fibre loss   5.1   dB
— fibre loss   81.5   dB
— fibre/receiver loss   0.17 dB
— Total   86.77 dB

The power available at the detector is

$$10 \cdot \log\left(\frac{f_r}{f_t}\right) = -87 \text{ dB} \tag{16a}$$

with:

$$\phi_t = 3 \text{ mW}$$

One has

$$\phi_r = 5.69 \times 10^{-6} \, \mu\text{W}$$

Equation (1) of the link assessment gives

$$10 \cdot \log\left(\frac{\phi_r}{\phi_t}\right) = 102 \text{ dB} \tag{16b}$$

It is thus possible to establish the margin of the link:

$$M = 10 \cdot \log\left(\frac{\text{sensitivity}}{\phi_r}\right) = 15 \text{ dB} \tag{17}$$

This margin of 15 dB enables account to be taken of ageing of the installation and the influence of temperature on the laser diode which is a critical parameter.

At the output of the pin diode, there is therefore a maximum of 2.45 μA which must be processed.

In the description of this link, two aspects have not been considered:

(a) The use of a transmitter–receiver pair over a very short distance. The assessment calculation must be repeated to determine the amplitude of automatic gain control required in order not to saturate the processing electronics.

(b) The type of signal — analogue or digital. In the case of an analogue signal, it is particularly important to examine the influence of the linearity of transmission and the minimum level of the analogue signal.

1.4. Digital transmission

Digital transmission involves transmission and reception of binary signals which are timed by a clock. Consequently, digital transmission must also recover the clock information of the signal to permit correct reading of the binary signals. The clock information could be transmitted separately but this would require duplication of the link; it is simpler to add it directly to the transmitted information and this is achieved by coding of the signal.

1.4.1. Coding of the transmission

Digital transmission on fibre is thus a transmission of coded data which ensures simultaneous transfer of binary and clock information; the operation of combining the signals is encoding.

1.4.2. Decoding

The receiving circuit must, in order to recover the original signal, provide two functions:

— a threshold detection function which permits discrimination between high levels (binary "1") and low levels (binary "0") of the received signal;
— clock recovery which enables correct reading of the transmitted binary sequences.

1.4.3. Error rate

The receiver must permit recovery of the binary sequences with a minimum of error; that is by distinguishing the 'high' and 'low' levels without ambiguity. This operation is disrupted by deterioration of the signal which arises mainly from the following:

— The limited bandwidth of the fibre which causes an overlapping of normally separate pulses (such as the binary sequence 101); there is then intersymbol interference.
— Spurious pulses, due to random noise, which generate unexpected binary 1s when their amplitude is greater than the detection threshold and their duration is compatible with the bandwidth.

The quality of transmission is defined by its error rate, which is determined by the ratio of the number of erroneous bits to the number of bits transmitted.

Transmission is considered to be good when the error rate is less than 10^{-9}; this usually implies a signal-to-noise ratio of 12, which must be included in the analysis of the digital link.

It should be noted that the processing electronics which follows the receiver must have a bandwidth equal to or greater than that of the fibre to avoid increasing the error rate.

It is generally assumed that the probability of a random error can be represented by the following function:

$$P_\varepsilon = 0.5 \cdot \text{erfc}(Q/2^{1/2}) \tag{18}$$

where

$$Q = \frac{1}{2} \cdot \frac{S}{N} \qquad \text{erfc}(x) = \frac{2}{\pi^{1/2}} \cdot \int_x^\infty \exp(-x^2)\, dx \tag{19}$$

and the usual values of error rate are given in Table 11.1.

1.4.4. Minimum optical power and error rate

By using signal-to-noise calculations for detectors (see Chapter 10), the minimum power can be related to a given error rate:

$$\phi_{0\min} = \frac{1}{S_\lambda \cdot m \cdot M} \cdot \left[2 \cdot I_{eq} \cdot \left(\frac{S}{N}\right) \cdot \Delta f \right]^{1/2} \tag{20}$$

This optical power increases with the bandwidth (Δf), the dark current, thermal agitation and the S/N ratio but decreases with the sensitivity (S_λ) and

Table 11.1

P_ε	10^{-6}	10^{-7}	10^{-8}	10^{-9}	10^{-10}
$\dfrac{S}{N}$	9.5	11	11.22	12	12.72

the avalanche factor (M). This minimum power clearly depends on the input impedance of the photodiode preamplifier.

By way of example, consider the case of a pin photodiode having the following characteristics:

sensitivity: $S_\lambda = $ 0.5 A/W
dark current: $I_0 = $ 1 nA
load resistance: $R_L = $ 1 MΩ
load capacitance: $C_L = $ 10 pF
modulation rate: $m = $ 2 (binary case)

The equivalent current I_{eq} has a value

$$I_{eq} = 2qI_0 + \frac{4kT}{R_L} \tag{21}$$

For a bandwidth of 100 MHz, the minimum optical power is as follows:

$$\phi_0 = 1.2 \times 10^{-8} \text{ W} \qquad S/N = 12$$
$$\phi_0 = 2.6 \times 10^{-8} \text{ W} \qquad S/N = 50$$

In practice, the results depend greatly on the amplifier circuits used; these are generally realized in FET or bipolar technology.

1.4.5. Transmission codes

(a) Selection criteria   The criteria involved in the choice of a transmission code include the following:

— the baseband spectrum, which must be as narrow as possible with respect to the bandwidth to limit detection noise;
— the capability of recovering the clock easily;
— the capability of transmitting any binary sequence;
— monitoring of the transmission error rate;
— ease of commissioning.

(b) Codes used   Numerous codes are used in conventional telecommunications, of which three are often applied to fibres

— the *non return to zero* (NRZ) code, where the 1 bit remains at 1 throughout the interval and the 0 bit remains at 0 throughout the interval;
— the *return to zero* (RZ) code where the 1 bit has the value 1 for the first half of the interval and 0 for the second half and the 0 bit has the value 0 throughout the interval;
— the *biphase S* (BI-S) code where the 1 bit makes a transition at the start

of the interval and the 0 bit makes a transition at the start and in the middle of the interval.

Of these three codes, the most common for fibres is NRZ which is used from low frequencies (kHz) to high frequencies (GHz) for both light-emitting diodes and laser diodes.

1.5. Analogue transmission

There are three classes of light-wave modulation which permit the transfer of analogue information as follows:

— intensity modulation;
— frequency modulation;
— pulse position modulation.

Each of these has a direct influence on the quality of transmission.

1.5.1. Intensity modulation or baseband modulation

The signal to be transmitted, possibly modified in amplitude, directly modulates the luminous intensity of the source. This simple form of modulation requires the least bandwidth since it is identical to that of the signal to be transmitted; in contrast, it is directly affected by the linearity of the component used.

(a) Non-linear distortion   Figure 11.2 shows the variation of luminous power emitted by a diode as a function of current.

The mean point about which intensity modulation is performed is chosen to be in a linear region and the modulation factor about this position is

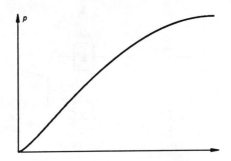

Figure 11.2   Luminous power transmitted as a function of the intensity of a light-emitting diode

defined as follows:

$$m = \frac{I_{max} - I_{min}}{I_{max} + I_{min}} = \frac{(2I_e)^{1/2}}{I_m} = \frac{P_{max} - P_{min}}{P_{max} + P_{min}} \tag{22}$$

where I_e is the effective modulation current. As an indication, a diode operated between 10 and 110 mA has a typical differential gain distortion of 6%.

To improve the linearity, one can either reduce the modulation factor or compensate for the non-linearity; there are several available methods, of which negative feedback is the most common.

This technique (see Fig. 11.3) is the simplest to use and consists of extracting a small portion of the transmitted power and modifying the control current.

By this process, the amplitude of the distortion is divided by a factor $(1 + \beta a)^2$ while the transmitted power is divided by $(1 + \beta a)$, where β defines the amplitude of the negative feedback and a is a factor of proportionality related to the electronics.

The problem is to optimize the factor β for a low harmonic content, good loop stability and high power.

(b) Signal-to-noise ratio and optical power   The determination of signal-to-noise ratio in Chapter 10 (Section 9.2) shows that it increases in proportion to the square of the modulation factor and consequently, the minimum optical power for a given signal-to-noise ratio is inversely proportional to the modulation factor.

$$\phi_{0min} = \frac{1}{S_\lambda \cdot m \cdot M} \cdot \left[2 \cdot I_{eq} \cdot \left(\frac{S}{N} \right) \cdot \Delta f \right]^{1/2} \tag{23}$$

This expression is similar to that for digital transmission except that the value of modulation factor has changed; the optical powers necessary are thus of the same order.

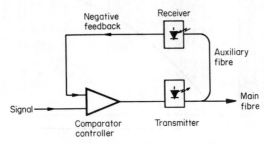

Figure 11.3   Linearization by negative feedback

1.5.2. Frequency modulation

The analogue signal is subjected to an amplitude–frequency conversion before transmission by the fibre; the inverse operation is performed on reception. This is a conventional method of improving the signal-to-noise ratio and decreasing the non-linear distortion of the transmitter; but this problem is transferred to the encoders and decoders. The main advantage of this technique is to permit frequency-division multiplexing of several channels.

The main disadvantage of this technique is the need for a greater bandwidth than in the case of baseband modulation. Equation (24) enables the required bandwidth to be calculated:

$$f_{FM} = 2 \cdot f \cdot (1 + \beta) \tag{24}$$

where f is the bandwidth of the signal and β is the modulation index which corresponds to the ratio of the frequency excursion to the maximum frequency.

For example, for a bandwidth of 6 MHz and $\beta = 1.5$, a bandwidth of 30 MHz is required.

Moreover, the frequency (F) of the sub-carrier must avoid interference between the second-order intermodulation products so that its spectrum occupies only an octave:

$$2(F - f_{FM}/2) \geqslant F + f_{FM}/2 \tag{25}$$

The system is thus centred about 45 MHz with the extreme amplitudes ranging from 30 to 60 MHz.

Calculation of the signal-to-noise ratio of the signal is related to that of the subcarrier $(S/N)_F$ as follows:

$$\left(\frac{S}{N}\right)^2 = 3\beta^2 \left(\frac{S}{N}\right)_{F^2} \tag{26}$$

which can be evaluated by conventional methods.

This method permits a gain of 15 dB with respect to baseband modulation and is indispensable for retransmission of television signals from satellite or radio links since these signals arrive already modulated by this method.

1.5.3. Pulse position modulation

The analogue signal is carried by a sequence of very short pulses (≈ 20 ns equivalent to 50 MHz) for which the interval between two pulses is proportional to the amplitude of the signal. This technique has two main advantages:

— It avoids the non-linearity of the transmitter.

— It leads to a low mean power dissipated by the optical source, which increases its lifetime.

The disadvantages are the complexity of the modulators and demodulators used and the need for a high bandwidth due to the shortness of the pulses. The signal-to-noise ratio is difficult to calculate since the necessary modelling of pulses depends on the pulse width; it is of the same order of magnitude as, or even an order of magnitude worse than, that of frequency modulation.

1.6. Modulation of optical signals — summary

Intensity modulation is invariably used with digital transmission which needs only limited linearity since the quality of the link depends essentially on the coding used. This technique is also used for analogue signals if the linearity criterion is not too severe.

Frequency modulation is by far the most used for analogue signals since it is a well-understood technique already used in other areas; adaptation to optical fibre links does not pose a problem.

In general, the advantage of avalanche diodes over conventional diodes increases as the signal-to-noise ratio decreases; it is negligible for signal-to-noise ratios greater than 50 dB.

1.7. Distribution of data

Data whether digital or analogue may require to be accessible to several users simultaneously. To achieve this, a data distribution network which is more complex than a point-to-point link is used. There are two principal network configurations; these are the serial network and the star network but combinations of the two are often used (hybrid networks).

1.7.1. Serial distribution

Serial distribution is a similar structure to a point-to-point link; all the stations are connected to the same fibre on which the data is transmitted. In the simplest configuration, the principal station transmits the data to all the stations on the line with no return of information. A more elaborate version permits each subscriber to send information to the principal station. Finally, for a true serial network, each station can transmit information to all other stations and receive it from all other subscribers. In the most commonly encountered case, the serial network is realized as a loop; that is all stations

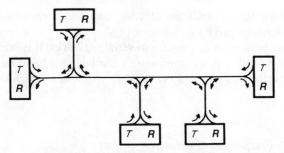

Figure 11.4   Configuration of a serial network

are connected to a loop of fibre which carries the information. A transmission protocol is then installed in each station to send or receive data on the fibre (for example by using the 'token' technique).

Coupling on to the optical fibre is made by means of four-port passive couplers. To increase the reliability of links, it is preferable not to use active couplers since these require an external electrical supply and are dependent on the reliability of the electronic components used.

Calculation of the link budget is performed as for a point-to-point link but account is also taken of the losses introduced by insertion of the passive couplers.

1.7.2. Star distribution

Star distribution is a means of increasing the number of subscribers while retaining satisfactory transmission quality. The system is distributed around a star coupler or distributor which receives the signal transmitted by any

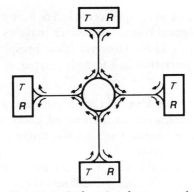

Figure 11.5   Configuration of a star network

station and forwards it to all the others. (The design and analysis of these couplers has been described in Chapter 7).

For the same reasons as in the case of serial networks, it is preferable to use passive couplers, although a compromise must be made between the technical difficulty of implementation and the reliability of a link which uses electronic components.

1.7.3. Comparison of structures

There is attenuation of the signal at each coupling of a serial network equipped with passive couplers. This requires that each receiver should have a sufficiently low noise level to receive a very weak signal, if it is far from the source, and automatic gain control if it is very close. This requires receivers of very high dynamic range to permit interchangeability on the network. The principal advantages of a serial network are its modularity, the length and number of sections of cable used. The star network has the advantage of sending a similar level of signal to each station while limiting the attenuation and propagation time of the line. The two major disadvantages of this network are the technical difficulty of realising passive couplers with a large number of channels and the quantity of cable required.

The choice of structure depends on the number of subscribers, their geographical situation and the operation of the network (whether interactive or not). A frequently used compromise is to install star networks in areas with a high concentration of subscribers and serial networks in areas where the stations are widespread; this is the basis of hybrid networks.

2. FIBRE OPTIC TRANSDUCERS

It is only since 1980 that fibre optic transducers have constituted a separate discipline but simple optical transducers which make use of light guides have been on the market since 1970. However, their prospective applications are immense and their exploitation has scarcely started; it is centred mainly on two types of use:

— The flexibility of the fibre and its robustness with respect to harsh external environments permit routeing of information to or from very diverse sites such as chemical and nuclear reactors, the human body and machines of various kinds.

— The characteristics of the light wave carried by the fibre depend on its form, its dimensions, stresses and deformation to which it is subjected, its temperature and so on. This enables many parameters to be used to modify the optical wave which propagates in the fibre and can thus be measured.

In accordance with these two categories of use, the transducer is defined as a system which contains one or more fibres which:

— carry information on the parameter to be measured;
— modify the optical wave as a function of the parameter to be measured.

This definition excludes the type of system which contains a conventional transducer and a fibre optic transmission system.

Fibre optic transducers are associated with two types of physical phenomenon:

— interferometry associated with phase or frequency modulation;
— amplitude modulation of the light wave (or the luminous intensity).

2.1. Interferometric optical transducers

Interferometric optical fibre transducers share the incident light between two fibres; one acts as a reference and the other as a sensor. The two returning waves are compared for any phase or frequency change caused by the environment on the sensor fibre.

The main, but not all, areas of application for these transducers are:

— fibre optic gyroscopes;
— acoustic sensors;
— sensors for electric and magnetic fields.

The Mach–Zehnder interferometer is the basis of all interferometric fibre optic transducers and particularly the fibre optic gyroscopes used for aeronautical guidance. The effect of the environment can be amplified by using

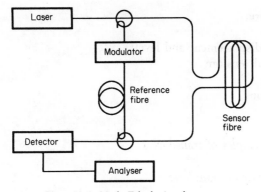

Figure 11.6   Mach–Zehnder interferometer

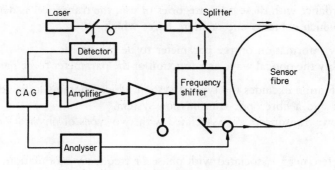

Figure 11.7 Fibre optic gyroscope

appropriate surface coating of the 'sensor' fibres. By way of example, the basic diagram of a fibre optic gyroscope is given above.

In this gyroscope, the incident beam is separated into two: one of the beams goes to the 'sensor' fibre and the other to a frequency shifter. The two beams are recombined at the output. In the absence of rotation, the phase relationship does not change. In contrast, the beam in the sensor fibre accelerates in the direction of rotation and decelerates in the opposite direction; this leads to a proportional frequency shift.

2.2. Modulating fibre optic transducers

Intensity modulating fibre optic transducers react to a variation of the luminous intensity within the fibre. This variation is caused by variation of the parameter to be measured. The perturbing environment can act on the light which enters the fibre or that which is flowing in the fibre. The main application areas of these transducers are:

— temperature;
— pressure;
— position, displacement and level;
— mechanical tension;
— vibration;
— acceleration.

2.3. Examples of transducers

2.3.1. Position transducer

This is the simplest and most commonly used transducer; a source emits light on to a fibre which transfers it to a detection system. An object, a gas or a

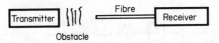

Figure 11.8   Position/displacement transducer

liquid passing between the source and the fibre causes a variation of intensity which is then detected.

This type of transducer can also emit light itself; this is reflected by the obstacle and returns to the detector by means of a return fibre.

This type of optical fibre transducer is generally used for the detection of position, presence of an object, displacement, level and so on.

2.3.2. Fibre optic temperature transducers

Numerous fibre optic temperature transducers have been examined in laboratories but few have been made commercially available. A wide range transducer (– 200 to 150°C) will be described here which has an accuracy of 2°C and was developed by M. C. Ferries of Kings College, London.

The transducer uses the temperature dependence of Raman diffraction in a fibre. This diffracted light is proportional to the Rayleigh diffraction in the fibre and the temperature.

$$I = A \cdot \frac{Ir}{\exp\left(\dfrac{h \cdot c \cdot V}{k \cdot T} - 1\right)} \tag{27}$$

where

I is the luminous intensity of the Raman diffraction,
I_r is the luminous intensity of the Rayleigh diffraction,
h is Planck's constant,
c is the velocity of light,
V is the variation in frequency associated with diffraction,
k is Boltzman's constant,
T is the temperature,
A is a constant of proportionality.

Figure 11.10 shows the principle of the transducer. The light wave from the

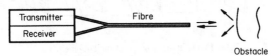

Figure 11.9   Position transducer using reflection

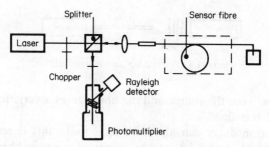

Figure 11.10 Fibre optic temperature transducer using Raman diffraction

laser is transmitted in a fibre which serves for transmission and reception. The received signal is sent to a Rayleigh detector and a photomultiplier; analysis of these two signals enables the temperature information to be extracted.

2.3.3. Fibre optic current and voltage transducers

An application of fibre optic transducers which is already highly developed is the measurement of electrical currents and voltages.

These measurements are made using the Pockel effect for voltages and the Faraday effect for currents; the fibre plays the role of information carrier.

(a) The Pockel effect A wave polarized in two orthogonal directions passes through a Pockel cell. At the output, the waves in the x and y directions have a phase difference which is directly proportional to the applied voltage. A quarter-wave plate is frequently used to increase the phase difference and increase the sensitivity of the phenomenon; the signal is received after passing through an analyser.

Typical materials used for Pockel cells have high electro-optic coefficients (Table 11.2).

(b) The Faraday effect A polarized wave passed through a Faraday cell has its

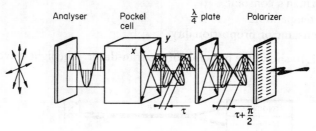

Figure 11.11 Basic diagram of the Pockel effect

Table 11.2

Material	Coefficient $(10^{-10}\,\mathrm{cm/V})$	λ (μm)
ZnSe	2	0.55
CdTe	6.8	1.06
LiNbO$_3$	8.6	0.55

direction of polarization rotated when a magnetic field is applied to the cell:

$$\phi = V_r \cdot H \cdot L \qquad (28)$$

where V_r is the Verdet constant, H is the magnetic field generated by the current in the material and L the length of the active region.

For this type of cell, materials with a very high Verdet constant are required; some of these are given in Table 11.3.

(c) The principle of a fibre optic current/voltage transducer   In this type of transducer, the fibre acts only as a light carrier.

The fibre transmits the light wave to the cell which is in a harsh environment and carries the information to the detector.

Transducers of this type have been tested in electrical generating stations and have given good results in comparison with conventional transducers since the fibre is unaffected by the harsh environment.

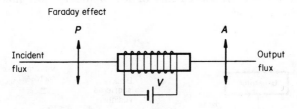

Figure 11.12   The principle of the Faraday cell

Table 11.3

Material	Verdet	λ (μm)
Flint	0.04	0.85
ZnSe	0.21	0.82
YIG	9.00	1.30

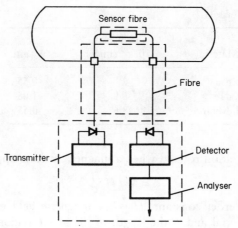

Figure 11.13 The principle of a fibre optic current–voltage transducer

2.3.4. Fibre optic vibration transducers

The Doppler effect laser velocity meter is a technique which is now used for contactless measurement of vibrating surfaces. The basic principle is to send a light wave on to a vibrating surface and analyse the scattered light to determine the frequency variation characteristic of the Doppler effect; this is directly proportional to the vibrations of the surface.

The fibre serves here both for light and information transport to and from the vibrating surface. Measurements of low velocities (0.2 m/s) have been made with this method.

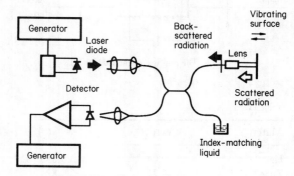

Figure 11.14 Fibre optic vibration transducers

3. IMAGING

Imaging with fibre optics is the least developed area of fibre applications in spite of numerous applications of medical endoscopy.

The principle is to transport the image by means of a bundle of fibres of which there can be several hundred. Each fibre carries the information relating to one pixel of the image; the information is received by a matrix detector of the CCD type which enables the image to be reconstructed.

Medical endoscopes have diameters of several tens of microns and provide both transmission of the light necessary to illuminate the observed region and return of the information. Their size enables them to be inserted in the human body through arteries, veins, tracheas and so on, thereby permitting direct observation of body cavities such as the heart, intestine, stomach and lungs.

BIBLIOGRAPHY

J.-Ph. Pérez, *Optique*, Masson, 1984.

G. Broussaud, *Optoélectronique*, Masson, 1974.

G. Bruhat, *Optique*, Masson, 1965.

L. Hedencourt and H. Lilien, *Optoélectronique*, Éditions Radio, 1983.

A. Cozonnet, J.-F. Curret, H. Maitre and M. Rousseau, *Optique et télécommunication*, Eyrolles, 1981.

Télécommunication Databook, 1984 NS.

M. Bruhat, *Cours de physique*, Masson, 1950.

J.-P. Mathieu, *Optique*, Vols. 1 and 2, SEDES, 1965.

A. Orszag and G. Hepner, *Les lasers et leur application*, Masson, 1980.

Optoelectronic applications manual: McGraw-Hill, 1977.

A. Hadni, *Essentials of modern physics applied to the study of infrared*, Pergamon Press, 1967.

Télécommunications optiques, Fibres multimodes, composants actifs, système, Masson, 1982.

C. Kittel, *Introduction à la physique de l'état solide*, Dunod, 1971.

F.-A. Jenkins and H.-E. White, *Fundamentals of Optics*, International Student Edition, 1981.

J.-P. Pocholle, Caractéristiques de la propagation guidée dans les fibres optiques monomodes, *Revue technique de la Thomson*, 15(4), December 1983.

M. Wher, S. Blaison and F. Gauthier, Fabrication des fibres optiques monomodes, *Revue technique de la Thomson*, 15(4), December 1983.

J. Puge and J.-P. Pocholle, Caractérisation des fibres optiques monomodes, *Revue technique de la Thomson*, 15(4), December 1983.

J.-P. Dehmichen, *Emploi rational des transistors*, Éditions Radio, 1980.

J. Bass, *Cours de mathématiques*, Masson, 1968.

L'avenir national des capteurs à fibres optiques: CIAME, 1984.

Fiber Optics Sensors — Proceedings, Vol. 568, 25 and 27/11/1985.

L'optique guidée monomode et ses applications, Masson, 1985.

J.-J. Clair, *Télécommunications optiques*, Introduction à l'optique intégrée, Masson, 1977.

R. Tiedeken, *Fiber optics and its applications*, The Focal Press, 1972.

Optoélectronique, Revue bimestrielle, Masson/ESI.

J.-P. Margeotte, *Interface intelligente pour réseau à fibre optique*, Rapport de stage DESS en informatique industrielle et optoélectronique, 1985.

INDEX

Absorbent substrate, 129
Absorption, 62, 156, 175
Absorption band, 63
Absorption efficiency, 133
Absorption loss, 52, 75, 78
Absorption spectrum, 5
Active coupler, 124
Active coupling, 109
ALPD process, 59
Amplifying medium, 155
Analogue transmission, 197, 207
Angle of incidence, 1
Angular misalignment, 114
Angular offset, 115
Anti-node, 4
APD, 185, 194
Attenuation, 5, 37, 62, 97
Attenuation coefficient, 90
Attenuation per unit length, 78
Avalanche factor, 190
Avalanche gain, 185
Avalanche noise, 188
Avalanche photodiode (APD), 185, 194
Axial lateral plasma deposition (ALPD) process, 59
Axial misalignment, 114
Azimuthal field, 11
Azimuthal mode, 166
Azimuthal order, 86
Azimuthal periodicity, 10

Backscatter factor, 90
Backscattering, 89
Backscattering measurement, 91
Bandwidth, 5, 37, 152, 194, 205
Baseband modulation, 207
Bessel function, 38, 47
BI-S code, 206

Bias delay, 169
Biphase S (BI-S) code, 206
Boltzman, 156
Boron, 56
Boundary condition, 11
Brewster angle, 1
Brillouin effect, 78
Brownian motion, 176
Buried cables, 106

Cable, fibre optic, 97, 101, 106
Cable filling coefficient, 104
Cable loss, 104
Cable parameters, 105
Cabling, 99
Calculated luminance, 137
Calibration, 72, 80
Calorimetry, 78
Carrier confinement, 164
Carrier mobility, 180
Cascade circuit, 148
Characterization of optical fibres, 67
Chopper, 124
Chromatic dispersion, 22, 25, 37
Cladding, 1, 56
Classification of modes, 42
Cleaving, 117
CMOS diode control, 148
Coding, 204
Coefficient of attenuation, 90
Coefficient of filling, 102
Coherence noise, 173
Collapsing, 57
Compact structure, 101
Conduction band, 125, 175
Confinement, 164
Confinement factor, 48
Connection loss, 109

Connection specification, 120
Connectors, 91, 120
Continuous mode, 142
Control photodiode, 170
Convexity, 116
Convolution, 83
Core, 1, 56
Core diameter, 110
Coupling, fibre–laser, 170
Coupling loss, 109, 122
Coupling of optical fibres, 109, 118
Crack, 98
Crack detection, 91
Critical angle, 132
Critical wavelength, 176
Current continuity equation, 169
Current limit, 148
Current regulation (diode), 146
Current transducer, 216
Cut-off frequency, 45, 84
Cut-off wavelength, 92

Dark current, 183, 194, 205
Decoding, 204
Defect, 30, 80
Degeneracy of modes, 47
Depletion region, 178, 180, 184
Deposition, 56
Detectivity, 190
Diffuser, 140
Diffusion, 180, 182
Diffusion loss, 52
Digital control of diode, 147
Digital transmission, 197, 204
Diode, 140, 142
Diode, control of, 148
Diode efficiency, 131, 134
Dispersion, 20, 26, 52
Dispersion loss, 75
Dispersion, modal, 10
Dopant, 25, 55, 62, 64
Doping 126
Doppler effect, 160, 163, 218
Double heterojunction, 128
Double heterostructure, 166
Drawing, fibre, 56, 61
Drift velocity, 181
Dynamic resistance, 145

Effective numerical aperture, 33, 68
Efficiency, 131

Efficiency, luminous, 136
Eigenvalue, 42, 47
Eigenvector, 42
Einstein constant, 157
Electric field, 8
Electric induction, 8
Electrical properties of diode, 142, 145
Electro-optic coefficient, 216
Electron density, 152
Electron spectroscopy for chemical
    analysis (ESCA), 65
Electron-hole pair, 178
Emission diagram, 141
Emission field, 130
Emission spectrum, 131, 141, 167
Endoscope, 219
Endurance, 98
Energy level, 125
ENP, 189
Epoxyacrylate resin, 62
Equivalent noise power (ENP), 189
Error probability, 205
Error rate, 204
ESCA, 65
Evanescent field, 11, 14
Evanescent mode, 13, 15
External quantum efficiency, 135
Extrinsic faults, 98

Fabrication of optical fibres, 55, 64
Fabry–Perot cavity, 159, 161, 164, 167
Fabry–Perot resonator, 155
Far field, 68
Faraday effect, 216
Fatigue, 98
Fermi band, 127
Fermi level, 127
Fibre link, 78
Fibre drawing, 56, 61
Fibre end, preparation of, 117
Fibre leader, 170
Fibre loss, 200
Fibre optic cable, 97
Fibre optic gyroscope, 213
Fibre optic link, 198
Fibre optic transducer, 212
Fibre parameter, 106
Fibre-laser coupling, 170
Field equation, 39
Fluorine, 56, 64
Flux, 137

Focal distance, 139.
Forbidden band, 125, 165, 175
Forward voltage (diode), 145
Frequency characteristic, 82
Frequency division multiplexing, 209
Frequency domain, 82
Frequency modulation, 209
Frequency response, 31, 82
Fresnel loss, 112, 131, 199
Fundamental mode, 166
Fusion, 119
Fusion-drawing, 122

Gaussian field approximation, 93
Gaussian impulse, 83
Germanium, 56, 64
Glass lathe, 56, 60
Graded index (GRIN), 3, 7, 77
Graded index fibre, 34, 56, 76, 90
GRIN, 7
Grooved cylindrical cable structure, 102
Group delay, 20, 29
Group propagation delay, 52
Group refractive index, 21
Group velocity, 90
Gyroscope, fibre optic, 213

Half amplitude width, 83
Halide, 65
He–Ne laser, 160
Head, laser, 170
Heisenberg's relation, 163
Heterojunction, 128
Heterostructure, 166
High-density injection 153
Homojunction, 126
Homostructure, 166
Humidity, 100
Hybrid mode, 44, 47
Hydroxide ion, 52
Hytrel, 99

Imaging, 219
Impedance of free space, 49
Impulse response, 31
Incidence, angle of, 1
Injection angle, 19
Injection condition, 19, 76, 78
Installation of cables, 106
Integrating sphere, 80
Intensity, 137

Intensity modulation, 207
Interface loss, 199
Interference field, 74
Interference microscope, 74
Interferometric method, 74
Interferometric transducer, 213
Intermodal analysis, selective, 88
Intermodal dispersion, 20, 22, 27, 84
Intramodal dispersion, 20, 84

Joule effect, 168
Junction capacitance, 181
Junction temperature, 142

Kevlar, 100

Lambertian source, 68, 138
Laser, 157
Laser amplification, 156
Laser diode, 155, 163, 173
Laser diode transmitter, 171
Laser effect, 157, 169
Laser equation, 158, 161
Laser head, 170
Laser noise, 173
Laser radiation, 162
Leader fibre, 170
Leakage current, 180
Leaky mode, 15, 70
LED, 125
Light guide, 1
Light-emitting diode (LED), 125, 154
Limiting angle, 1, 132
Linear diode control, 149
Linearly polarized mode, 48
Link assessment, 203
Link specification, 202
Load resistance (diode), 145
Logarithmic circuit, 192
Longitudinal field, 9
Loss, 75, 80, 97, 110, 199
Loss factor, 81
Loss per unit length, 76
Low-density injection, 153
Lumen, 136
Luminance, 137
Luminescence, 134
Luminous efficiency, 136, 145
Luminous flux, 138, 141, 149, 175
Luminous intensity, 137, 139, 144, 161
Luminous power, 136

Mach–Zehnder interferometer, 213
Magnetic field, 8
Magnetic induction, 8
Magnification, 139
Margin, 200
Material, 56, 62, 165
Maxwell's equations, 8, 45, 159
McDonald function, 38
MCVD method, 56
Mean diode current, 143
Mean luminous intensity, 144
Mechanical behaviour of cables, 105
Mechanical reinforcement, 99
Microfracture, 97, 99
Microlens, 141
Minimum optical power, 205
Minority carrier density, 152
Misalignment, 116
Mismatch loss, 199
Mismatching, 18
Mobility, 180
Modal dispersion, 10
Modal excitation, selective, 88
Mode, 10
Mode coupling, 30, 76
Mode dispersion, 21
Mode group, 18, 85
Mode jumping noise, 173
Mode number, 12
Mode order, 86
Mode, reinforcement of, 161
Mode scrambler, 77
Mode separation, 20
Mode width, 92, 94
Modes, classification of, 42
Modified chemical vapour deposit
  (MCVD) method, 56
Modulating transducer, 214
Modulation, 172, 207, 209
Modulation depth, 189
Modulation factor, 208
Modulation frequency, 169
Modulation index, 209
Modulus of transfer function (MTF),
  32
Monomode optical fibre, 4, 37, 39, 51,
  54, 81, 92
Monomode propagation, 48
MTF, 32
Multichannel cable, 102
Multidielectric structures, 53

Multimode optical fibre, 4, 7, 34, 85
Multiplication factor, 186

Near field, 68
Negative feedback, 208
Network, 211
Neumann function, 38
Node, 4
Noise factor, 188
Noise level, 194
Non return to zero (NRZ) code, 206
Non-linear distortion, 207
Non-linear effect, 78
Non-parallelism misalignment, 115
Non-radiative transition, 126
Normalized propagation parameter, 51
Normalized frequency, 42
Normalized mode, 18, 86
Normalized radius, 7
NRZ code, 206
Number of modes, 16, 18, 29
Numerical aperture, 2, 67, 76, 110,
  141
Numerical aperture, effective, 33, 68
Numerical aperture loss, 199

OH radical, 62
Operating limit of diode, 142
Optical cable, 101, 104
Optical coupler, 31
Optical efficiency, 131, 134
Optical fibre, 1, 140
Optical fibre, fabrication of, 55
Optical head, 170
Optical power, 208
Optical power, stabilization of, 172
Optical telecommunications, 197
Optical window diameter, 141
Optimum refractive index profile, 23
Oscillating field, 14
Output power, 154
Outside vapour phase deposit (OVPD)
  method, 58
OVPD method, 58

p–n junction, 126
Package temperature, 142
Parabolic profile, 8
Parallel connection (diode), 148
Passive coupler, 121
Passive coupling, 109

PCVD method, 58
Peak diode current, 143
Performance of optical cables, 104
Periodic bending, 30
Periodic constriction, 31
Periodic curvature, 30
Permeability, 8
Permittivity, 8
Phase condition, 160
Phosphorus, 56
Photocurrent, 175
Photodarlington, 191
Photodetector, 178
Photodiode with amplifier, 192
Photon, 157, 175
Photon confinement, 164
Photon lifetime, 160
Photon-electron conversion, 175
Phototransistor, 191
Pin photodiode, 181, 183
Planck's constant, 126
Plasma chemical vapour deposit
   (PCVD) method, 58
Plasma modified chemical vapour
   deposit (PMVCD) method, 57
Plastic fibre, 34
PMVCD method, 57
Pockel effect, 216
Point-to-point link, 210
Polishing, 118, 123
Population inversion, 157, 162
Position transducer, 214
Potential barrier, 127
Power consumption, 154
Poynting vector, 48
Preform, 55
Preparation of fibre end, 117
Primary protection, 99
Propagating mode, 13
Propagation, 65
Propagation constant, 9, 13, 18, 21,
   30, 51
Protection of fibres, 99
Protection of laser diode, 171
Pulse position modulation, 209
Pulse spreading, 20, 83, 85
Pulsed mode, 142, 169
Pumping, 162

Quantum efficiency, 134, 177

Quantum noise, 188

Radial field, 11
Radial misalignment, 113
Radial mode, 16
Radiant intensity, 137
Radiation condition, 162
Radiation diagram, 34, 68, 138
Radiative recombination, 134, 152
Radiative transition, 126
Raman diffraction, 215
Raman effect, 78
Rate of recombination, 134
Rayleigh diffraction, 215
Rayleigh scattering, 52, 62, 80
Receiver, 175
Recombination, 134, 153, 179
Reference fibre, 72
Reflection coefficient, 132
Refracted near field, 70
Refracted ray, 1
Refractive index, 1, 55
Refractive index dispersion parameter,
   25
Refractive index gradient, 12, 199
Refractive index parameter, 27
Refractive index profile, 3, 7, 12, 20,
   25, 65
Refractive index profile, dispersion of,
   21
Refractive index profile, measurement
   of, 68
Refractive index profile, optimum, 23
Refractive index profile parameter, 7
Refractive index profile, variation of,
   26
Refractive index profile, W-type, 54
Refractive index variation, 111
Reinforcement of mode, 161
Rejection loss, 122
Relative luminosity, 136
Resonant cavity, 159
Response, speed of, 169
Response time, 153, 180
Return to zero (RZ) code, 206
Ribbon cable structure, 102
Ronchi chart, 94
Ronchi mask, 93
Roughness, 116
RZ code, 206

Scalar wave equation, 37
Scattering coefficient, 80
Scattering loss, 80
Schottky noise, 188
Sealing, 100
Selective excitation, 87
Selective intermodal analysis, 88
Selective modal excitation, 88
Sellnier expansion, 22
Semiconductor, 125, 163
Sensitivity, 177, 194
Sensitivity of the eye, 136
Serial distribution, 210
Serial network, 121
Series connection (diode), 148
Shannon's theorem, 170
Sheet cable structure, 102
Signal-to-noise ratio, 85, 189, 194, 208
Silica, 22, 52, 98
Silicone, 72
Single heterojunction, 128
Single mode fibre, 4
Single-fibre cable, 101
Snell's law, 1, 19, 132
Space-charge region, 127
Spatial filtering, 77
Spectral attenuation, 52, 64, 81
Spectral density, 156
Spectral distribution, 157
Spectral filtering, 77
Spectral response, 178
Speed, detector, 180
Speed of response, 169
Splices, 91, 119
Spontaneous emission, 156
Spot diameter, 76
Spreading, pulse, 20
Spurious pulse, 205
Stabilization of optical power, 172
Star distribution, 211
Starck effect, 163
Stationary wave, 4
Step index, 3, 8, 34, 39, 77
Stimulated emission, 155
Strain, 97
Structural defect, 63
Substrate, 129
Surface area loss, 199
Synchronous detection, 77

Tape, 100

TE mode, 42
Temperature transducer, 214
Tensile strength, 98
Thermal agitation, 205
Thermal behaviour of cables, 105
Thermal noise, 188
Thermal properties of diode, 142
Thin section method, 74
Threshold current, 164, 168, 181
Time domain, 82
TM mode, 42
Total insertion loss, 122
Total internal reflection, 2
Transducer, 212
Transfer function, 32, 83
Transit time, 181
Transition, 126
Transmission, 198
Transmission code, 206
Transmission coefficient, 132
Transmission loss, 122
Transparent substrate, 130
Transverse electric mode, 43
Transverse field, 9
Transverse interference method, 75
Transverse magnetic mode, 42
Transverse mode, 46
Triangular profile, 8
TTL diode control, 148

VAD process, 59
Valence band, 125, 175
Vapour axial deposit (VAD) process, 59
Vapour phase deposition process, 28
Variable attenuator, 124
Verdet constant, 217
Vibration transducer, 218
Voltage transducer, 216

W-type refractive index profile, 54
Wave equation, 8, 11, 37
Wave function, 11
Wavelength, critical, 176
Wavelength multiplexer, 124
WBKJ method, 10
Welding, 119

Young's modulus, 99

Index compiled by J. C. C. Nelson